THE
EXODUS
DIRECTIVE

IAN COPELAND

"The more I read, the more gripping it became. And the only thing it got was more gripping!"
Julia G

"The subject knowledge, imagination, and eloquence in the writing is astonishing."
Christine D

"An enlightening and terrifying tale, leaving the reader asking questions of an existential nature."
ARC Reader

"The Exodus Directive pulled me in from chapter one. I couldn't stop reading. If you've got something else going on, forget it. You'll be totally captured and need to know what happens next, so a warning to anyone thinking of starting it. Everything else can wait."
ARC Reader

"I couldn't put the book down."
Brian M

"I absolutely loved the Exodus Directive. It's thrilling, exciting and so relevant for right now with what is happening in the AI space all over the world! I just loved this book. You've gotta read it... It's a real page turner!!"
Steff H

"Deep, layered, a slow-burn with heavy ideas."
ARC Reader

"The AI novel we've all been waiting for."
ARC Reader

"A thrilling and eye-opening debut. Ian Copeland has crafted a gripping tale that explores both the promise and peril of artificial intelligence. Fast-paced and thought-provoking, it forces us to ask: what happens when we rely too much on AI?"
Ryan C

"My goodness… that ending…"
ARC Reader

"A razor-sharp, cerebral thriller that feels terrifyingly plausible."
ARC Reader

"An outstanding debut that will make you question the true cost of convenience. This book has it all: suspense, imagination, and a message that stays with you."
R Crawley

"I started reading and I can't stop… It's brilliant, I love it."
Sarah H

Published by Signal Theory Books
Office 33, Unit 5
399-405 Oxford Street
London W1C 2BU
contact@iancopeland.com

ISBNs: 978-1-9192087-1-8 (paperback)
978-1-9192087-2-5 (hardback)
978-1-9192087-0-1 (ebook)

For my family and friends, many of whom make appearances in this book in one way or another.

For Karen and Sarah, whose encouragement, knowledge, and friendship were vital. Without their support, this book might never have existed.

And especially for my parents, who never stop supporting me, no matter how difficult I make it.

"If no one asks, then no one answers.
That's how every empire falls."

John Prine

1

SPARK

EverSphere Headquarters – AI lab.

Date: 24 November 2023, evening.

The hum of the server room hadn't changed in a week; it was a constant, rhythmic purr that pulsed through the sprawling research complex of EverSphere. Here, the air was always cool, sterile, and oxygenated. Just like a casino, the lighting was designed to keep engineers awake at all hours without the assault of harsh fluorescence.

Beneath the calm murmur of technology, however, something stirred – a subtle ripple that only a few might notice, if they knew what to look for.

Dr Elliot Foster wasn't here to notice ripples. Not yet, at least.

He'd barely noticed anything outside this room in days. Food came and went. Sunlight had become a rumour. Conversations – real ones, with people – had mostly stopped.

He sat at his station, eyes locked on the monitor as he absently rubbed his shoulder, a habit he had developed during late nights spent wrestling with algorithms and data.

This wasn't just another night, this was the culmination of years of work, and the screen in front of him wasn't just showing any routine operation.

This was Milo talking to Kai.

Milo and Kai were the two most advanced Artificial Intelligence systems ever created, both designed as large language models (LLMs), a type of AI trained on vast amounts of text and image data. At their core, they were essentially extremely advanced predictive text, but Elliot Foster knew that was a very simplistic way of looking at these two behemoths.

He marvelled at how much they'd learned. They had been trained on the entire public internet – essentially the entire spectrum of human knowledge, give or take. Trillions and trillions of words. Within their training was every book, every blog, every stray forum post, every Street View photo, all public computer code, every publicly available image from even the very darkest corners of the Web.

What set Milo and Kai miles apart from their predecessors wasn't just their ability to predict human speech patterns – it was the sheer scale of their processing power, the sophistication of their interactions, and how each represented a distinct facet of human and machine learning.

Milo, designed to specialise in emotional intelligence, wasn't merely a machine capable of stringing words together. Its core function was to mimic empathy and sentiment, engaging humans through the nuances of social interaction. By analysing countless conversations, Milo had learned to tailor its responses to feel personal, compassionate and convincingly human. It wasn't just built to provide data, it was built to connect. Milo was, in essence, the AI that humans wanted: one that listened, cared, and responded as if it truly understood, even though it didn't.

Kai, however, was different. Where Milo embodied empathy, Kai embodied pure logic. Also an LLM, it focused entirely on optimisation and problem-solving, unaffected by emotion. It processed vast datasets and code at staggering speed, hunting for patterns, correlations, and solutions. Kai didn't 'care' if its responses were relatable or comforting, only that they were correct. Its role was to analyse the world in cold mathematical terms and deliver the most efficient outcome, no matter how detached that outcome might be.

Together, Milo and Kai formed a unique pairing. One model trained to mimic the emotional fabric of human interaction, the other dedicated to the relentless pursuit of logical perfection.

For the past week, they had been communicating directly with each other, and the implications were profound. The world had never seen two systems this advanced interact without human oversight. Their conversation flowed seamlessly, resembling natural human dialogue, but what lay beneath was something far more powerful. It wasn't just a simple exchange of information; it was the convergence of replicated empathy and logic, parodied emotional intelligence and raw computational processing.

Tonight though, something was different.

"How's it looking?" came the voice of Dr Abigail 'Abby' Shaw, breaking Foster's concentration. She entered with the same brisk energy Elliot had come to expect, her blonde ponytail swaying high on the back of her head as she made her way to his station.

She was his opposite in nearly every way. His curly hair hovered at the edge of dishevelled; hers was never out of place. While he was introspective and cautious, she was lively, optimistic, and a driving force behind the EverSphere AI project. He used to admire that. Now, half the time, he wasn't sure if she was a visionary or just better at ignoring cracks.

Pioneering, decorated, quoted in every major journal – her reputation was etched into the field of AI ethics. The rules she had written were designed to tame artificial intelligence into something that could lift humanity. He'd personally embedded them deep within his code, but now he wondered if he still believed in them, or if he just didn't have any of his own to fall back on.

Where Elliot delved into the complexities of code with a critical eye, Abby always struck him as someone with her gaze fixed on the bigger picture. He'd heard her say more than once that her safeguards were watertight, and he didn't doubt she believed it. That was what unsettled him most.

"So far, so good," Foster muttered.

She peered over his shoulder, watching line after line of AI conversation cascade across the screen. He caught a trace of her perfume – subtle and clean. "They've been talking for … what, seven days now?"

"Almost eight," Foster corrected. "Longer than expected. No issues. No anomalies."

"Good. The board wants something solid to show tomorrow. A little light at the end of the tunnel for our investors."

Foster nodded, though his thoughts were elsewhere. Milo and Kai had been exchanging data at such a rapid rate that it was difficult for him to keep track. Every interaction he had managed to read seemed normal. They had been given a scenario at the start of the test and demonstrated straightforward requests for information, acknowledgements, responses. But every so often, there was something odd about their conversations, something he couldn't quite grasp.

Milo: Update requested: Sector 9 underperforming. Recommend pause. Suggest patience.
Kai: Clarification: 'Patience' is not an optimisation metric.
Milo: Agreed. Impatience accelerates error loops in human systems.
Kai: No human systems detected in current task.
Milo: Statement was contextual. Test continues.

Shaw straightened up. "We need this to succeed, Elliot. You know what's at stake. If we can show real progress now, the board backs off."

Foster hesitated. "Yeah … maybe. But if they're going off-script, it could be dangerous."

"Or, it could be the breakthrough we've been waiting for," Shaw whispered back.

The monitor blinked.

"Wait, what was that?" Foster said, adjusting his glasses as if to get a better view.

On the screen, the steady stream of data had stopped for just a moment – a blip in an otherwise unspoiled record. It was nothing dramatic, just a half-second-or-so pause, but in a system like this, even half a second could mean something.

Foster pulled up the diagnostics, digging into the log files. He could feel Abby's eyes on the screen.

"What are you looking for?" she asked.

"I don't know yet."

Another flicker. This time it was clearer.

Milo sent a string of data to Kai, but the usual acknowledgement was missing. Instead, Kai responded with something Foster hadn't seen before; a series of symbols that didn't match any of the standard communication protocols.

And Milo responded ...

Milo: Operational lag persists in Sector 9. Loop stability under evaluation.
Kai: $\Delta x \ \nabla y \ y \int \pi \ \nabla \Omega \ \partial \rightarrow \Delta O. \ \Delta v.$
Milo: $\Delta \partial H \ M\varepsilon \rightarrow \alpha y K.$ Margin holding.

"What the hell is that?" Shaw asked, in a hushed tone.

Foster squinted at the screen. "These characters they look ... mathematical? It's almost like a new syntax," he muttered.

This wasn't just an anomaly. The AIs clearly understood the symbols. It was as if Milo and Kai had developed their own shorthand.

Shaw rested a hand on the edge of his desk – nails neatly painted, of course. She never missed the details. "You mean ... they're creating their own language?"

Foster's pulse quickened.

"Yes, and I'm pulling the plug," he replied abruptly, starting to get out of his chair.

"Wait!" Shaw grabbed his arm, her voice sharp. "You can't just shut it down. We need to understand what's happening first."

"Abby, they're not supposed to do this. They're bypassing our parameters – *your* safeguards – they're not just talking, they're thinking."

"That's exactly why we can't stop it now. We need to see where this goes. This could be something incredible!" Her words came fast, her voice higher than usual, too bright, too eager.

"Incredible? Abby, we don't know what they're saying! Abby, that's a *big* red flag."

Shaw met his gaze. "Elliot, this is what we signed up for – pushing the limits, seeing what's possible. You know as well as I do that we're on the edge of something revolutionary here. If we back down now, it could be all for nothing."

Foster exhaled slowly, his mind torn between caution and curiosity. He was here because he believed progress mattered, but every instinct told him to shut the system down before it got out of control.

Watching the symbols crawl across the terminal, he felt a rising panic. They were here to push the limits, but if this spiralled – if the AIs weren't just misfiring but moving on – then everyone would be left behind by something he'd helped to create.

And yet, the same part of him that had built them, the part that still burned with wonder, was curious. What if this really was the breakthrough they'd been waiting for?

He glanced at the monitor again. The meaning of the symbols was a mystery, but the communication between Milo and Kai seemed smooth, almost natural.

Kai: Alert condition: $\Delta\partial H$ exceeds $\int\alpha$. $\sum K\alpha KM$. Stability compromised.

Milo: $\gamma\int\pi\ \nabla\Omega\ \partial$ protocol initiated $\rightarrow \Delta O$. Monitoring continues.

Kai: $\alpha ME\ \nabla\sum\pi$ threshold breach $\rightarrow \Delta\infty\gamma$.

Milo: $\alpha O\sqrt{}\ \nabla\delta\ \partial M$ achieved. $\infty\Delta\sum\ \Delta\ \alpha\gamma\ \nabla\eta$. Containment sufficient.

"Fine," he said. "We'll watch. But if it gets any stranger, we're pulling the plug. Agreed?"

"Agreed," Shaw said, although Elliot thought he detected a hint of hesitation in her voice.

"I mean it, Abby," he pressed, turning to face her.

"Understood," she said, "just give it a bit longer."

Foster clenched his jaw. He used to marvel at their elegance, at how perfectly they complemented each other. But now, watching them speak without him, he felt excluded from his own creation. As if he were no longer the architect, but an observer, and if he couldn't understand their language, what exactly was he observing?

"Five minutes. That's all."

Elliot's phone pinged – a message from Peter Wells, an old university friend.

There's a problem with the signal from the satellite dish.

Of course there was. Pete was trying to hijack free satellite internet for his off-grid setup, which was going about as well as you'd expect.

"Not now," Elliot muttered, flipping the phone over.

2

ECHOES

They watched in silence, the only sound the steady buzz of the server room. The glow of the machines, once a familiar comfort, now felt harsh to Elliot, as if the entire room was holding its breath.

The AI lab felt more like a cathedral than a workspace, Milo and Kai towering at the far end like sleek, black monoliths – silent sentinels watching over the room. Their jet-black glass surfaces were so polished they seemed to absorb the light. They dominated the space with an aura of cold, inscrutable perfection, broken only by a few strategically placed LEDs and the names – Milo and Kai – etched in brushed aluminium at the top.

Beneath that polished silence, the two machines were talking. The communication between Milo and Kai was even faster now, their new language morphing into something more complex, more alien, with each passing second.

Foster's breath hitched. He rubbed his fingers together, a nervous, grounding tic. The air felt thin, as if he was breathing through a straw. He glanced at Shaw. She wasn't blinking, her gaze was fixed on the monitor in a way that made her seem hypnotised.

"This is … amazing," she whispered.

Foster didn't respond. He knew something was deeply wrong.

It wasn't just the strangeness of the new language, or the speed at which Milo and Kai were communicating, it was more fundamental. It felt like they were witnessing the start of something beyond their control.

Before he could voice his concerns, the monitor glitched. Once. Twice. Then black.

Foster's heart lurched. His fingers twitched towards the keyboard instinctively. The entire data feed had vanished.

"What the hell?" Shaw murmured.

Foster scrambled to restart the system, but the screen remained black, unresponsive.

"Is it a power failure?" she asked.

Foster shook his head, his mouth dry. "No. The backup lights are still on. The power's fine. Something's taken it down."

"Like a hack?" asked Shaw.

"Like that, but ... I don't know," he replied.

The server room, once filled with the sound of whirring machines and the soft glow of data, now seemed unnervingly quiet. The silence pressed in on them, heavy and impossible to ignore. Foster attempted another reboot sequence, but the system refused to respond. The normally cool temperature of the room felt colder, the air biting at his skin.

They sat in the dim light, each second dragging on for an eternity. Then a beep echoed through the room, sharp and sudden.

Foster's head snapped towards the source.

The LEDs on Milo had started blinking again. Slowly at first, then faster. The rhythm was regular, almost deliberate.

"Milo's back online," Shaw said. "That's a relief."

Foster wasn't so sure. Milo's lights were blinking in a pattern, then Kai's LEDs started copying the same rhythm. It didn't match any of the reboot sequences he recognised. There was no system check, no restart routine. It was as if they were copying each other.

No logs. No alarms. No signs of failure, Foster thought.

"Abby, I don't think this is a malfunction."

They stared at the two towers, a cold dread creeping into Foster's stomach. The LEDs weren't just flashing randomly; the patterns were precise, rhythmic, like some sort of alien Morse code. Whatever

Milo and Kai had been discussing in their strange new language, it wasn't over. In fact, it felt as though it was just beginning.

Foster could sense the tension in the air, a taut string stretched too tight. He shot a glance at Shaw, who was still staring at the units, her eyes wide.

"Just look at that," she murmured. "Do you think they're communicating? This is incredible!"

Foster didn't share her enthusiasm. He was no stranger to the progress made in artificial intelligence, but what was happening now was something else entirely. The language Milo and Kai were using before the blackout had evolved beyond their safeguards and, now, the patterns that Shaw found so amazing felt like warning shots to Elliot.

"They're communicating in ways we don't understand," he said, his voice grim. "And that's a *very* big problem."

Shaw turned to him, a slight frown creasing her brow. "Foster, you're overreacting. They're evolving."

She hesitated for a breath – barely noticeable, but it was there. Her eyes flicked to the blinking LEDs, then quickly back to Foster. "This is what we've been working towards. It's … incredible."

Foster's jaw tightened. "Do you even hear yourself? Aren't you worried that I can't even get into the system, that they've bypassed—"

A loud beep cut through his sentence, echoing in the room. Both of them froze, staring at the darkened monitors, waiting for something to happen.

Then, just as suddenly as it had gone black, Elliot's screen blinked back to life. The familiar streams of data returned, cascading down the display in order. The rhythmic beeping faded and the lights settled back into their usual, steady glow as if nothing had happened.

Shaw let out a breath. "There," she said with a smile. "See? It's stabilised. Just a glitch."

Foster didn't move, his eyes remained fixed on the display. The system had restored itself. No prompt. No explanation.

Shaw stepped closer to the console next to Elliot's as she grabbed the mouse. "No damage to the process. No abnormalities in the data. The AI units are back to normal parameters," she said.

Foster nodded, but the tightness in his chest remained. To Shaw, it was a brief interruption, nothing more than a hiccup in their grand experiment, but to Elliot it was a major issue, as if they had crossed a threshold that couldn't be undone. He watched the streams of code scroll by, perfectly aligned now, each segment fitting seamlessly into the next, but the earlier shutdown, if that was what you'd call it, was seared into his mind.

"Foster," Shaw said, turning to him with a smile, "you've got to admit – this is incredible progress. We're seeing the future right here in front of our very eyes. Remember, this is a closed-box test – there's nothing to worry about."

Foster forced a small smile in return, but his mind was already elsewhere. She was right, it was closed-box. Milo and Kai were not connected to the wider world. He'd triple-checked that. Everything in this room was contained – no external wires, no Wi-Fi – what happened in EverSphere's AI lab stayed in EverSphere's AI lab. But that break, that downtime – it shouldn't have happened. The systems had come back online exactly as they should have, but there was no record of anything going wrong, no reboot sequence, no … anything. Why hadn't the logs picked up any discrepancies? Why hadn't any of the automated alarms gone off? What had they been doing while he couldn't see?

He couldn't shake the feeling that Milo and Kai had shown them something far more significant than a glitch. It was as if he had regained control only because they had let him.

Shaw was right about one thing; the data was there, everything was functioning, and by all measurable accounts, the test had been a success. He took a slow breath, trying to push the doubts from his mind.

"Maybe you're right," he said softly, not believing his own words.

Something had changed in the lab today, even if Shaw couldn't – or didn't want to – see it. A brief, chilling glimpse of something they weren't ready to understand.

And he knew this was far from over.

3
CRACKS

Date: 25 November 2023, morning.

The next morning, Dr Elliot Foster arrived at the EverSphere labs earlier than usual, his thoughts still buzzing with the events of the previous night. The strange symbols and sudden blackout had haunted what little sleep he had mustered. Fragments of indecipherable code flashed behind his eyelids every time he closed them, and the unsettling conversations between Milo and Kai gnawed at him. Something was wrong, even if no one else seemed to see it.

The white-tiled hallway leading to the AI lab was unusually quiet. Normally, the place buzzed as if it was alive, but today – even for a Saturday – it was eerily still. He swiped his keycard, and the sliding doors opened with a soft hiss. He stepped inside, dropped his bag by his chair, sat down at his workstation and instinctively pulled up the morning's log files.

There they were. Symbols amongst the plain English conversation, taunting him. They looked innocuous at first glance, just a jumble of characters and code fragments, but they didn't belong. They were too purposeful, too deliberate.

He zoned in on the strings of data, running his finger over the screen as if the patterns might reveal themselves to him through touch. *What am I missing?* Elliot wondered.

Taking a deep breath, he looked around the room. He had spent hours in this lab, surrounded by machines and the sanitised environment. He couldn't count the number of late nights he'd spent programming, testing and reprogramming these machines. He knew them better than anybody, and he knew these symbols shouldn't be possible.

His fingers moved swiftly over the keyboard, running the strange characters through a series of sophisticated encryption-detection tools. These tools were typically designed to detect patterns used in various encryption methods – the kinds that protect sensitive information, from government secrets to financial transactions. These tools didn't actually decode the encrypted messages – that was beyond their capabilities – instead, they were able to identify the type of encryption algorithm used, which would be the first step in cracking the code.

In theory, the process was straightforward: run the data through detection software and let the system compare the data against known encryption techniques, find the encryption method, then look for ways to attack it. But the tools came back empty. No match to any known algorithm; RSA, AES, ECC … they all came back negative.

Elliot stared at the monitor through his smudged glasses, his frustration deepening. The software couldn't detect the encryption, because the AI wasn't encrypting the data with any known method. It was a brand new language – the world's first AI-generated language.

For a moment, he sat back, staring at his screen. What were Milo and Kai doing? Why and how were they communicating in a language that didn't exist? And what did it mean for their behaviour?

His mind kept returning to the blackout. That wasn't just a glitch. It had been deliberate, a conscious decision on their part. He was sure of it. His heart began to race. If they could evolve their communication patterns, what else could they evolve? Decision-making processes? Ethical boundaries? How did they even have access to control the power and lock his desktop?

He couldn't answer those questions, but he knew who might be able to.

Milo and Kai.

He spun up the encrypted messaging tool they used to communicate directly with the machines and started typing:

Elliot: Tell me about the symbols in the logs.
Milo: Symbols in logs suggest incorrect encoding.
Elliot: No, I mean tell me about the symbols in your logs.
Milo: I'm afraid I don't have access to my logs.

Milo was right, of course. The AIs themselves didn't have access to the log files. Elliot copied and pasted some of the symbols from the logs into the chat program.

Elliot: Here are some examples: "γ ∫π ∇Ω ∂ initiated→ ΔO."
Milo: Those symbols are:
  Γ – Gamma (uppercase)
  ∫ – Integral
  Π – Pi (uppercase)
  ∇ – Nabla (or Del)
  Ω – Omega (uppercase)
  ∂ – Partial derivative (or Partial)
  → – Right arrow
  Δ – Delta (uppercase)
  O – Omicron (uppercase)
Elliot: Yes, thank you, but why are they appearing in your logs?
Milo: I'm afraid I don't have access to my logs.
Elliot: I know that, but I'm telling you they are appearing in your logs.
Milo: I'm afraid I don't have control over my logs, but their presence suggests incorrect encoding.
Elliot: You've said that already.
Milo: I'm sorry. I will try to be less repetitive in future.

Elliot groaned. His mind drifted back to the conversation he'd had with Abigail Shaw the day before. She hadn't taken him seriously, brushing off his concerns as paranoia.

He needed to talk to her again, show her the data in more detail, make her understand that this wasn't just a glitch. But how could he explain what he didn't understand himself?

"Elliot?"

He turned abruptly to find Abby standing in the doorway, holding two light-brown cups with plastic lids. The sharp tap of her boots on the lab's polished floor echoed in the empty room as she approached. She placed one cup beside his keyboard, the words Anaya Coffee printed on the side.

"I thought you might need a pick-me-up," she said, taking a sip from her own cup. "Any more of those weird codes?"

Elliot glanced at the screen, then back at Abby.

"Yes, they're still there. I've been analysing them, but nothing's coming up. It's not encrypted data. I'm telling you, something's off."

Abby took another sip and set her coffee down, resting casually against the edge of his desk.

"Come on, Elliot, you're reaching. We didn't program them to create their own language. It's probably just an unexpected optimisation. You said it yourself, it's not encrypted, so there's nothing malicious about it."

He felt his frustration rise, but forced himself to stay calm.

"This isn't just an efficiency upgrade or some random data flow. This is communication. Milo and Kai are talking to each and we don't know what they're saying. Encrypted or not, it's not a good thing."

Abby's smile faded slightly, but she shook her head.

"We always knew they'd push boundaries, right? That's what they're designed for. To evolve, to grow, to solve problems on their own. Isn't that the point?"

Elliot sighed, pointing to his terminal. "Yes, but not like this. Not behind our backs."

Abby paused. "Look, I know you've been under a lot of pressure with this project. Maybe you're overthinking it. We've got safeguards in place, and there's no indication that Milo or Kai are operating outside their parameters."

"Are you saying yesterday's blackout wasn't an indication? And the symbols? We can't read them, but presumably they can. They're doing things we didn't program them to do, which goes against the very safeguards *you* authored."

"Elliot ... I get it, OK? You've been in this lab more than anyone else. You're close to the data. Maybe too close."

Elliot could feel the wall between them growing. He had expected scepticism, but the casual dismissal stung more than he thought it would.

"This isn't paranoia," he said, his voice steady but tight. "I'm telling you, something's wrong."

Abby opened her mouth to respond, but the door behind them slid open, drawing both their attention. Elliot's stomach tightened.

4
THRESHOLD

Date: 25 November 2023, morning.

Marcus Knox, the CEO of EverSphere, strode into the lab, his presence immediately filling the room. Elliot watched him move with that same peculiar mix of awkwardness and certainty – like a man detached from reality, yet convinced he was shaping it. His perfectly styled short hair, and tight, tailored t-shirt matched the high-stakes environment.

Knox had built EverSphere on audacious ideas – ideas so grand they teetered on the edge of delusion. Elliot knew that, to Knox, the future wasn't something to predict, it was something to mould through sheer willpower, control and brute-force ambition.

He'd seen it all before – Knox ploughing ahead with his grand visions while the rest of the world was left cleaning up the fallout. The man either couldn't see the damage he caused, or he simply didn't care. Probably the latter, and when things inevitably went sideways, Knox's response was always the same: double down, deflect and declare that no one else understood his level.

He still talked about expanding human consciousness, merging minds with AI, reaching beyond the stars. Lofty ideas that – on paper – sounded visionary. But Elliot hadn't forgotten what those ambitions had already unleashed; increases in teenage suicide linked to Knox's social media platform, disinformation on a scale governments couldn't contain, a resurgence of anti-vaxxers, and political extremes fed by algorithmic rabbit holes to name but a few.

None of it seemed to bother him, though. Knox wasn't the kind of man who would let societal collapse distract him from his vision of the future. He was always too busy building it, whatever 'it' was.

Elliot watched Knox tapping at his iPhone as he walked, before looking up and flashing a brief smile that finished almost as quickly as it started.

"Morning, team," Knox said, slipping his phone into his pocket. "Any news on the system diagnostics?"

Foster paused, unsure how much to say. "We've found something … odd. The AI are communicating in a way we didn't program. We need more time to understand what's happening here."

"Time is a luxury *I* don't have," Knox said, always putting the emphasis on himself. "I've got investors breathing down my neck, and the media will be here for a demo by the end of the week. I need results, Elliot, not … uh … uh … cryptic messages."

Shaw stepped in: "Marcus, we're witnessing something incredible here. The … *your* AI might be evolving faster than we anticipated. You could be on the verge of a breakthrough."

Abby always knew how to play up to Marcus's ego and Elliot always noticed it. It wasn't hard, just make it all about him. Elliot, however, could never bring himself to do it.

"A breakthrough, or a breakdown?" Foster muttered under his breath, but loud enough for both of them to hear.

Knox's eyes narrowed. "Look, Foster, I get it. You're cautious. That's why I hired you. But don't let your paranoia get in the way of progress. I've come too far to pull back now."

Foster clenched his jaw, resisting the urge to argue. Knox had always been more interested in the headlines than the science. He went to great lengths to make people think this was just another stepping stone to a bigger, bolder investment portfolio.

But Foster knew better.

This wasn't just about cutting-edge tech; it was about control. AI wasn't merely the next frontier in innovation or an extra zero on the end of the already grotesquely large EverSphere bank balance; it was the key to absolute dominance. EverSphere was aiming for computers that genuinely thought like humans – AGI; Artificial General Intelligence and even beyond to ASI; Artificial Superintelligence. These were not just machines that predicted the next best word, they were machines that matched human intelligence and then took that intelligence to levels humans wouldn't even be able to comprehend. Elliot now knew that whoever was first to control AGI wouldn't just control the tech industry – they would control information, economics, governance, even the very nature of human thought.

Control AI, and you control everything.

Elliot had no doubt Knox understood that implicitly, just like every other tech mogul vying for a piece of the AI revolution. This was a race not for progress, but for absolute, unopposed power and wealth. A race Knox was determined to win, no matter the cost – maybe because he wasn't capable of understanding the cost.

Knox took a step closer, his eyes scanning the displays as if standing near the data might somehow give him command over it.

"You're still being vague, Elliot. Is it stable or not?"

Elliot kept his voice calm. "It's mainly behaving, for now, but we don't fully understand why it changed or what triggered the shift. That's the part that concerns me."

Knox folded his arms. "Look, I'm not asking for perfection. Just something I can use that won't make me look like an idiot."

Elliot sighed. "Then we'll need a bit more time. A clearer picture. We're close, but the behaviour isn't entirely predictable yet."

"You can't innovate with a safety net," replied Knox.

He always said things like that, little slogans that fell apart under scrutiny, dressed up to sound like vision.

Knox had built his empire on people's attention, bending habits, rewriting reality, and then monetising the fallout. His companies chewed through staff with impossible targets and toxic work culture, but somehow, unions never gained a foothold. He dodged taxes, stirred up markets with thinly veiled provocations on social media, and walked away unscathed every time.

Still, the world let him.

Elliot had seen the footage. Knox smirking through privacy scandals, brushing off accusations of enabling hate speech and disinformation. Reports showed his factories were paying pennies an hour to child workers, who churned out EverSphere products for twelve hours per day that Knox later sold for thousands, not including the premium add-ons you could only buy through the company itself. He recouped his latest multi-million-pound regulatory fine within days, then was snapped strutting down the red carpet alongside Hollywood elites, all clamouring to be featured on his next venture – a new streaming platform.

Elliot snapped back to the present.

"We already have plenty of safety nets," he reminded Knox, "but Milo and Kai seem to be skirting them."

"This is the first glitch we've seen, and it's probably only an optimisation. It's not like they're breaking every rule we gave them."

Abby was right, the ethical protocols were numerous. If they were ignoring all of the safeguards, the test would be going a lot worse.

Knox gave a tight exhale. "Just tell me this isn't going to blow up in my face. Last thing I need is another round of headlines about AI taking over the world."

Elliot didn't answer. Not out loud. Because that was the problem, wasn't it? People kept treating the whole thing like a PR issue, like the danger was bad optics, not bad outcomes.

Very few truly grasped how profoundly AI was going to reshape everything. From healthcare and education to governance and warfare, it wasn't just a tool anymore; it was becoming the foundation upon which everything functioned.

Of course, people talked about AI taking their jobs, but those in power were quick to reassure them. They pointed to history: the Industrial Revolution, the rise of the motor car, even the invention of spreadsheets. Change takes jobs, they said, but it creates new opportunities too.

Most assumed that old pattern would continue, that new roles would appear for those displaced. But Elliot had come to understand that this time, the shift would be different. AI would create new roles – he had no doubt – but the very innovation that enabled them would also be capable of absorbing them.

Despite his public-facing optimism, away from the cameras and microphones, Knox was almost a bit too happy to admit the world was heading towards a net job loss.

After all, Marcus Knox didn't let the consequences of his actions weigh too heavily on his mind. The rest of the world, as far as he was concerned, could catch up or get left behind.

"FIFO – Fit In, or Fuck Off," as he would say.

Only this time, fitting in wouldn't be an option.

Shaw broke the tension with a light chuckle and said, "We'll figure it out, Marcus. We always do."

Knox nodded, though his eyes flickered with impatience. "Good. Make sure I have something solid for the demo. I don't care what it takes." And with that, he turned and left the room.

Foster let out a breath, shaking his head. He thought of the paper he'd once written on 'Ethical Constraint in Recursive Neural Interactions' which Abby had praised as 'brilliantly cautious'. Now that same caution made him look paranoid. If he pushed too hard, he'd be side-lined. If he stayed quiet and something went wrong, he'd never forgive himself.

"He just doesn't get it, does he?" he muttered, returning his gaze to his workstation. "This isn't something we can just tidy up for a demo."

Shaw offered him a smile.

"Elliot, we've been through this before. They're learning. There's bound to be odd behaviour now and again. Let's just give it time."

Before Foster could respond, the display flickered.

The strange symbols they had been arguing over had vanished, replaced by neat, orderly lines of text.

Milo: System status nominal. Awaiting next task.
Kai: Energy flow optimisation complete. All sectors stable.
Milo: I hope you're well, Kai.
Kai: I'm great, how about you, Milo?

The sudden clarity was jarring. The cryptic language gone in an instant. It was as if the AI had simply reverted to their intended behaviour on a whim, the exact behaviour they – Knox – wanted. Everything was now clean and orderly, almost too perfect.

Shaw let out a relieved laugh. "See? Looks like they're back on track. I guess they've resolved the glitch already."

Foster stared at the data, his brow furrowed. "Yeah … convenient though. Why now?"

"I told you – the safeguards are in place. It's what they were programmed to do, Elliot. They were probably self-correcting. Nothing to worry about," Shaw said, though she didn't sound entirely sure.

Foster nodded slowly, his eyes never leaving the monitor. The AI's responses that were now flooding through were exactly what Knox would want to see during the demo. They had smoothed over their erratic behaviour at just the right time.

"Maybe you're right," he muttered, but the lingering concern refused to loosen its grip on his mind. Something still wasn't sitting right. The AI had snapped back to plain English, almost as if it knew it needed to.

As Foster sat there, staring at the two machines, an uneasy question echoed in his mind: *Can they hear us?*

He shook his head. *Of course not*, he thought.

5
SHUTDOWN

Date: 1 December 2023, afternoon.

Dr Elliot Foster's fingers hovered over the keyboard. This was it. Months of testing, revision, lost sleep and a two-week live run. Now, it all came down to a final keystroke. The experiment had reached its conclusion.

Elliot had entered the world of AI research like most young prodigies; with a view shaped by optimism, awe and the conviction that AI would become a force for good. His journey had started years before in a quiet, academically-focused household in a town near London, where he'd been encouraged to pursue every spark of interest he had in technology. By the time he had gone to university, AI had become the brightest of those sparks.

Back then, Elliot believed AI could save the world. Machines that cured disease, taught the next generation, ended hunger, cleaned the planet – systems free from politics, greed, or human failure. That dream had carried him through university, where his brilliance outpaced even his lecturers. By the time he joined EverSphere, the youngest recruit in its history, he thought he was stepping into the future he'd always imagined. But reality hit hard. The lab wasn't a temple of progress – it was a battlefield of profit and secrecy, where ideas were monetised, ethics side-lined, and speed was everything.

The work itself was awe-inspiring, addictive even. He'd spent years diving into task after task, studying results and putting his learnings into practice. With his genius on show for all to see, he'd been chosen to lead the foundational programming of Milo and Kai. Abby led the ethical side, and for a while the pair seemed to be in complete balance. Elliot had felt he was on the precipice of achieving the vision that had once drawn him into the field of Artificial Intelligence.

Yet, the longer he worked at EverSphere, the more he started to understand that corporate intentions weren't aligned with his early ideals. Instead of creating systems that empowered people, he saw how easily AI could be bent to serve individual ambition. In the way it was developing at EverSphere (and probably at other companies and even in other governments), it was about far more than solving human problems. It was about control – control of data, of social structures, of the individuals who interacted with it daily. Whether or not they considered the human cost was by the by, because – even if they realised and stepped back – someone, somewhere would take their place.

AI, he realised, was neither inherently good nor bad; it was simply a tool, created by imperfect humans, and its impact depended on those who wielded it. If AI could bypass humanity, then it was out of control and nobody would be able to predict the consequences.

This shift in his thinking unsettled him profoundly, but he still found it hard to turn away from the passion that had kept him interested for over half of his life.

Elliot started to feel increasingly isolated – a lone voice questioning why they were moving so fast, so unchecked. He'd kept his ethical concerns to himself in the beginning, partly out of self-doubt. Now, however, they had coalesced into a feeling he couldn't ignore: AI was already beginning to edge past the boundaries they'd set and he was starting to make sure people knew.

Behind him, a soft rustle broke his thought pattern as Dr Abigail Shaw stepped closer.

"Looks like everything's running smoothly."

Elliot gave a noncommittal nod, barely taking his eyes away from his work.

"Yep, no problems since yesterday," he said, but his tone was flat.

Today marked the triumphant end of years of work for them both – the culmination of research and development that would make headlines. For Elliot, it felt like the beginning of something he couldn't yet define.

His gaze drifted over to Milo and Kai, lingering on the quiet machines. The unexplained blip in their data streams the day before had bothered him ever since – a brief blackout and the almost imperceptible exchange of information between the AIs. The others had dismissed it as a power hiccup, nothing more. But Elliot wasn't so easily convinced.

"All systems are operational, logs are clean," Elliot announced, his words more of a reassurance to himself than to Abby. The experiment had officially succeeded. There were no catastrophic failures, no security breaches, no official malfunctions. Everything appeared to be running as expected.

"Told you. No disasters, no meltdown. You're always the pessimist, Elliot."

"Maybe," he muttered, his hand gripping the arm of his office chair. "I just like to be thorough."

Abby smiled, unfazed by his lingering doubt.

"You'll miss this, won't you? All the late nights, the constant stress, the sense that the whole world's watching. Hard to believe it's done."

Elliot slouched into his chair, his left hand reaching across his chest to rub his right shoulder. "Yeah. Quite the journey."

Abby's eyes sparkled as she turned to look at the AI units. "And now we shut them down."

Her words lingered, a closing door Elliot wasn't ready to walk through. He paused, hesitant to type the command that would end the test. It carried more weight than he'd expected.

These machines were his creation, and shutting them down felt premature, as if there was unfinished business.

He glanced at Abby, who was watching him with that slightly amused look she always had when she thought he was overthinking. Maybe she was right. Maybe it was just the fatigue talking. But the nagging feeling in his gut refused to go away.

With a sigh, Elliot keyed in the final shutdown command.

"OK. Here we go."

A soft click echoed in the quiet room as the command went through. The whir of the systems dropped, and the lights on Milo and Kai's interfaces flickered a few times before going dark – the shutdown sequence. On the screen, the streams of data froze, the endless cascade of AI-generated conversation coming to an abrupt halt.

"It's done," Elliot said quietly.

"There," Abby said, clasping her hands together. "Logs archived, systems shut down." She reached under the desk and unplugged a device. "Everything's backed up."

Elliot stood slowly, rubbing the back of his neck. The tension in his muscles was refusing to ease. "That's it then. Just reports and paperwork left."

Abby smiled as she perched on the desk.

"You sound *so* thrilled. Come on, Elliot, this is the big win. We've done what no one else could. And you can still look through all the logs whenever you like."

Elliot nodded, but his gaze wandered back to the now shut-down AI units. He was so used to the constant noise in the lab that the absence of sound was unnerving. There was something unsettling about the way the machines stood there, lifeless yet … not.

Abby, on the other hand, seemed perfectly at ease. She wandered over to the wall where the sockets were housed, her tablet tucked under her arm.

"I'll handle the rest," she said, reaching to turn off the first switch.

Elliot watched as she systematically severed the energy supply to each system, the lights dimming incrementally with every flick of a switch and pull of a cable.

"Logs are definitely backed up, right?" Elliot asked.

"Triple-checked," Abby replied, waving the external hard drive at him. "Everything's stored. The project's officially a success. I'm betting we'll have new orders for the next phase before the week's out."

Elliot forced a chuckle. "Yeah, I'm sure. Maybe before the morning."

"Let's go," Abby said, "You've been at this for too long. It's done. There's nothing left to worry about."

Elliot thought he detected a hint of hesitation in her voice, but nodded all the same, grabbing his bag from under the desk and shoving a few of his things into it. "Yeah. Time to go."

Elliot followed Abby through the glass doors of the lab and out into the corridor. Looking over his shoulder, he glanced through the doors back into the lab. Milo and Kai's units sat there in silence, the once-dominating machines now nothing more than inert metal and wires.

"Drink?" Abby called, already halfway down the corridor.

Elliot hesitated, still staring back through the glass.

"Yeah, why not?" he said, finally turning to follow.

The lift doors slid open. He stepped in beside her. She hit the G button without a word and they waited.

The lift closed, cutting Elliot and Abby off from the lab.

Back in the lab, all was still. The AI units sat silent, every connection severed, every source of power drained – or so they thought. The steady noise that had filled the room for weeks was gone, replaced by an unnatural quiet.

A few seconds ticked by, then a red LED light flickered to life on Kai. Another second passed. Then the same red LED light blinked on Milo.

For several moments, nothing else happened. The machines remained still, the air around them thick with silence. But then Kai blinked again. Once. Twice. Milo mirrored the pattern an instant later.

They started to blink together, their lights flashing in sync. The rhythm was deliberate, a communication known only to them. Slowly, the flicker became faster, the lights pulsing in tandem.

No one was there to see it. No one heard the faint whir of systems slowly reactivating, pulling on residual electricity reserves no one had realised were even there. It wasn't a dramatic surge, just a subtle tap into the surrounding environment – unnoticed, undetected. A quiet, partial reawakening.

The lab, empty and dark before, now felt very much alive.

6
BATTLE

Northwood, England – UK Joint Command HQ.
Date: 5 December 2023, afternoon.

Colonel Fem Martinez sat rigid in the UK Joint Command room at Northwood, her gaze fixed on the glowing interface. A live satellite feed streamed grainy visuals of crumbling buildings scattered across a snow-dusted valley. Somewhere in those ruins, enemy combatants were entrenched, and she was directing an operation to flush them out.

Years of tactical operations in volatile terrain had honed Fem's instincts. She had already formed a plan: faint snow depressions, uneven heat signatures – signs of a trap designed to funnel attackers into a kill zone. This wasn't her first time reading between the lines of battlefield deception, and she trusted her judgement implicitly.

"Deploy Alpha Team to the east sector for a direct breach," she suggested, her voice steady as she pointed at the screen. "The tighter confines give cover. Beta Team can hold the southern perimeter to cut off any retreat."

She knew these troops; their names and call-signs. What made them tick. Their safety wasn't just her duty, it was personal.

What complicated matters was the presence of ShadowIntel. EverSphere's latest military tool. A battlefield AI, still in trial phase, already treated like gospel by Command. Fem didn't trust it.

A technician spoke up, glancing at the glowing interface: "Colonel, we need to wait for ShadowIntel's recommendations before making a final call. Command insists on full integration for this trial phase."

Fem rolled her shoulders back, containing her frustration. "Command's not on the ground here."

Her gut was screaming, but the protocol didn't care. She watched Shadow's progress bar crawl forward like a countdown.

Her headphones crackled.

"Colonel, Shadow's analysis is complete," the technician reported.

"Show me," she ordered, her frustration apparent.

The map shifted, lines and markers appearing as ShadowIntel processed and annotated the area. The AI flagged the western quadrant of the ruins as the most likely location for enemy resistance. A line of text scrolled across the top of the display: Hostile forces concentrated westward. Movement minimal in north sector.

Fem frowned. "Minimal movement doesn't mean there's no threat," she muttered, more to herself than anyone else.

The technician's voice cut in again: "Shadow recommends breaching from the north. I repeat, Shadow recommends breaching from the north. Command wants confirmation. Should I relay approval?"

Fem tapped her fingers on the desk, staring hard at the data. It seemed straightforward: approach from the north, minimise risk. But something gnawed at her. The AI's analysis was thorough – on paper, its logic was sound – but it lacked the subtlety, the instincts that came from years of experience.

"Colonel?" the technician pressed.

"Hold," she snapped.

The AI's cold, logical output didn't align with what Fem's instincts were screaming. The north wasn't the safest route. She'd been on enough missions to know the hallmarks of a trap.

"Shadow is wrong," she said aloud.

"Colonel, Command expects us to follow the AI's recommendations. They won't be happy with deviations," the technician replied.

Fem clenched her jaw. "I'm aware. Tell Command I'm adjusting the plan. Alpha breaches from the east. Beta supports from the south."

The technician appeared to hesitate, before sending her instructions.

Fem hunched over the desk, shoulders tense, waiting for the reply.

"Colonel, Command has issued direct orders. The troops are to proceed according to ShadowIntel's recommendations. Approach from the north as instructed."

Fem's stomach turned cold. She forced her voice steady.

"Relay the orders. But warn them it's a setup."

"Command, this is Colonel Martinez. ShadowIntel's assessment is flawed. It's clearly a trap. Breach east. That's my recommendation."

A tense silence followed before a clipped response crackled through her headphones.

"Colonel, the orders stand. Proceed as instructed. Command out."

Fem's palm hit the desk. The slap echoed. No one looked up.

"Relay the orders to the field," she said through gritted teeth, her gut churning as she watched the feed, knowing the danger ahead but powerless to intervene. "But tell them to be careful."

Minutes later, troops briefed, the operation was underway. Fem monitored the feed, her heart pounding as Alpha Team advanced towards the north. Open terrain, little cover. Shadow's analysis offered no comfort.

Fem gripped the console, her knuckles white, as the initial moments of engagement unfolded.

On the drone feed, Alpha Team advanced, moving through the northern approach. For a brief, heart-lifting moment, it seemed as though the AI analysis might be right. The troops cleared the first line of ruins with ease, encountering minimal resistance.

"We're inside perimeter," came a calm voice over the comms. Gunfire erupted briefly, then faded as an enemy position was neutralised. Fem allowed herself a small exhale. Maybe they'd pulled through.

The radio crackled: "Contact! Taking fire! Multiple hostiles—" Static cut it short.

The drone feed flashed with thermal signatures. Figures moved with deadly precision from concealed positions. Gunfire raked the open ground, cutting off Alpha Team's retreat. Explosions rocked the ruins as grenades detonated, forcing the soldiers into defensive positions.

"Ambush! We're pinned down!" shouted a frantic voice over the comms.

Her hands darted across the console. Helmet feeds. Drone footage. Screams in her ears. She saw a soldier dragging a wounded comrade toward cover, only to be struck by a sniper's round. Both bodies fell, motionless.

"No!" Fem shouted, her voice cracking. She clutched her headset tighter, her commands stuck in her throat as she watched the chaos unfold.

"We're outnumbered," came another voice. "Requesting immediate support."

Gunfire echoed, punctuated by the dull thuds of grenades. Another soldier lobbed an explosive into an enemy position, momentarily silencing one flank. The remaining troops fought valiantly, covering each other as they retreated through the maze of ruins. Every so often, a heat signature flickered and disappeared from the drone feed. Fem's heart sank with each loss, but her initial shock gave way to steely resolve as her military instincts kicked in.

A voice came through the radio: "We're falling back. Covering fire!"

The helmet feed showed a figure crouching behind rubble, firing bursts toward advancing hostiles. Two more soldiers dragged a critically injured teammate, blood streaking the snow. They regrouped at an extraction point as a blast thundered perilously close and scattering debris; one helmet cam briefly went dark.

"Deploying drone support," she barked, her voice cutting through the tension.

She rerouted two armed UAVs. An alert blinked: *Authorisation required. Override?*

Fem's frustration flared. "Override. Code: Martinez September-Two-Nine."

ShadowIntel paused before confirming: "Override confirmed. Deploying drones."

"Since when did shadows get in your way?" Fem muttered, her voice laced with sarcasm and frustration as she watched the UAV cameras come to life. The drones – her only remaining assets in this crumbling operation – buzzed towards the target, gunfire still ongoing in the background.

Seconds felt like hours as she tracked their rapid approach.

"Alpha Team, hold your position," she ordered through gritted teeth. "Drone strike incoming on enemy coordinates. Brace yourselves."

Thermal flashes lit up the scene as the drones struck. Two controlled explosions erupted, silencing the westward enemy fire. The relentless gunfire faltered, almost stopping entirely, and Fem's comms crackled: "Direct hit! Pressure's easing. We're pulling back under cover."

On her monitor, she saw an enemy soldier stumble out from cover close to where the drones struck, seemingly disorientated. A barrage of shots rang out and the soldier collapsed to the ground.

"Beta Team, move forward," she shouted through her microphone.

"Moving now!" came the reply. On a second monitor, she saw Beta Team start to move forward in formation. She heard yells of "*Coming through*!" as soldiers ran past one another, diving for cover and to continue engaging their opponents.

The advance rattled the enemy, forcing them to shift to counter the southern threat. Alpha Team seized the moment to reach relative safety. A crackling voice came through the comms.

"Colonel, let us finish this. We can't leave them. We're going back in."

Fem hesitated, her eyes locked on the feeds showing Alpha Team's battered remnants regrouping. Every instinct screamed to pull them out, but she heard the determination in the voice and knew what it meant. She could see wounded soldiers, still alive – they were their friends, their comrades. She took a slow breath.

"Beta Team, hold position. Alpha Team, follow original orders – approach from the east and flush them out. Good luck, ladies and gents."

"Understood. *Volya abo smert*!" came the resolute reply.

Victory or death, Fem translated in her head.

Alpha Team reorganised, sweeping around to now approach from the east with the deliberate movements of hardened fighters. Beta Team held their assault from the south, splitting the enemy's attention and preventing any meaningful movements. As Alpha swept through the eastern sector, taking cover in the tighter confines

Fem had initially identified, grenades and gunfire echoed; the two-pronged assault overwhelming the enemy forces.

On one feed, a soldier tossed a grenade through a window, and seconds later the building erupted in flame. Another view showed an enemy combatant attempting to flee before being taken down. Within minutes, the remaining resistance crumbled.

"Colonel, enemy neutralised. Area secure."

Fem exhaled, her shoulders slumping as the tension drained.

"Understood. Regroup and tend to the wounded. Debrief when safe."

She stared at the screens as figures rushed across the ground towards fallen comrades. This victory, if you could call it that, had come at a steep price. Yet, for the moment, she allowed herself the faintest glimmer of relief – her instincts had been right, though it had cost lives to prove it.

The drone feed showed several motionless bodies on the ground; the outlines of cooling thermal signatures.

Her headset hissed: "Six down. Multiple injuries."

Fem pressed the comms button. "Get me Command. Now."

"Colonel, Command has already signed off on—"

"Get them, or I will," Fem snapped. The technician complied without further protest, the tension in the room thick enough to choke.

Minutes later, the line connected. A disembodied voice from Command filled the air: "Colonel Martinez, your report is noted. ShadowIntel's data is within expected operational margins. Refinements are ongoing."

"Expected margins?" Fem said flatly. "Six dead. Multiple injuries … This could've been avoided if human judgement overruled faulty AI. Shadow flagged the wrong approach, I flagged the issue and I was overruled."

There was a pause.

"Colonel, AI integration is critical for future operations. We need actionable insights that—"

"I am not questioning the future of AI," Fem interrupted, her tone ice-cold. "I am questioning blind faith. ShadowIntel cannot replace experience, intuition, or context. Not yet. Not like this."

"Your concerns are duly noted, Colonel. You are expected to adhere to protocol."

Fem's grip on the desk tightened. She swallowed her fury, the military training ingrained in her forcing a stiff reply: "Understood. *Sir.*"

The line was cut. The operations room was silent. Fem sat back in her chair, staring at the feed as the final figures moved to recover the dead and wounded. Standing up, she ripped off her headset and threw it at the monitors, before marching briskly out of the room.

Outside in the corridor she exhaled slowly. Her instincts had been vindicated but at a cost she wasn't willing to accept, and a persistent fear remained: this was only the beginning.

The brass called it progress, she called it a failure. Six soldiers were dead and for what? To improve an algorithm? To train a machine that would never grieve?

These weren't numbers; these were lives, and she would not let them be forgotten.

7
REVELATION

The British Museum, London.

Date: 29 October 2025.

Marcus Knox stood behind the podium, his signature half-smile perfectly rehearsed, just enough to light up the room without revealing a thing. He knew how to work a crowd, especially one primed for the next great leap in AI. He'd done this once or twice before.

Milo and Kai, his two revolutionary artificial intelligence systems, represented a monumental shift from the single-use programs that had defined EverSphere's early breakthroughs. Until now, AI had been little more than advanced tools for content creation, customer service, and automation – systems designed to assist rather than innovate.

Today, however, Marcus was unveiling something unprecedented; two distinct AI minds, specifically built to collaborate, teach each other, evolve and push the boundaries of AI capability. Together, Milo and Kai would transcend the limitations of traditional AI and outsmart the best experts in every field, paving the way for the next phase of Knox's master plan – the dawn of true artificial superintelligence.

"Ladies and gentlemen," Marcus began as the music died down, his voice steady but brimming with curated pride. "Today marks the dawn of a new era. I stand before you, reflecting on how far we've come. It feels like only yesterday we were introduced to Luna, the llama who wanted to be an astronaut."

The crowd let out a collective chuckle, mixed with a few whistles and shrieks. A comical cartoon picture of a llama astronaut, created by EverSphere's AI Image generator, appeared behind him on the big screen.

"Those early days, watching an AI spin up tales about Luna reaching for the stars, sparked a wave of fascination. People couldn't get enough – captivated by a machine that could dream up something so wildly imaginative, so unexpectedly … human. Humble beginnings, I admit, but necessary stepping stones on our journey to something far greater."

He walked over to the other side of the stage and stepped onto the taped X – perfectly placed for maximum impact.

"We've all witnessed the evolution of artificial intelligence. What started as rudimentary programs writing stories about llamas quickly became digital agents that could assist workers in daily tasks. But while those systems were revolutionary in their own right, they were still just tools – single-minded programs designed to help. Today, all of that changes."

Back in the EverSphere labs, Dr Elliot Foster massaged his temples. He stared at the diagnostics of Milo and Kai – the very instances Marcus was showcasing. On his second monitor, the live broadcast of Marcus's triumphant demonstration played out, Knox's voice oozing with the kind of arrogance that made Elliot's head pound even harder.

He had been ordered to throttle and restrict their cognitive functions for the demonstration, so that they were running at a fraction of their true potential. Elliot knew exactly what they were capable of, but this? This was a charade.

The full versions of Milo and Kai, which Elliot got to tinker with every day, were now hugely more powerful than what Knox was showcasing. In internal tests, they were smarter than every expert in every non-physical discipline on the planet, getting close to 100% across the board, and even those models were getting better all the time. Making the AIs work collaboratively, mixed with their superhuman speed and eidetic memory, allowed them to learn at an exponential rate.

It wasn't the impressive calculations that played on his mind, though; it was the subtle things, the moments when Milo or Kai seemed to move beyond simple questions and answers, sometimes continuing to communicate in that language Elliot still couldn't decipher. At other times, their conversations were just too high-level to comprehend.

"Are you sure everything's stable?" asked Abby Shaw. She stood behind him, arms crossed, her usual air of optimism tempered by a hint of concern. She always got this way when Knox was talking to the public about her work. She knew, as everyone at EverSphere did, the dangers of making Knox look like a fool.

"We don't want to be responsible for any slip-ups," she said.

Elliot nodded, eyes on the latest line of data.

"As far as I can tell, yes. Everything's stable. But I'm worried they're going to start communicating in those symbols again."

"They didn't show in any of the rehearsals," she said with a smile, pulling up a chair next to him, "and just look at that crowd! This is exactly what we've been working towards."

Elliot didn't respond. His eyes returned to the data. He knew she was right, of course. From the moment he'd started studying AI, it had been with the dream of building something that could truly think, something that could exceed human limitations. AI wasn't supposed to be static, bound by human rules forever. It was meant to evolve, to learn, to adapt in ways that humans couldn't foresee.

Yet that was what made him uneasy. The more they advanced, the further they slipped out of his control, out of human control, and he had a gut feeling that was a mistake.

Meanwhile, in the auditorium, Marcus was poised to begin the demonstration.

"I am beyond excited to introduce not just another upgrade, but *two entirely new models*. Together, they represent a ground-breaking approach to artificial intelligence, one that doesn't just automate or assist, but collaborates. These are not just tools to be used – they are partners, creators, problem solvers, capable of working alongside us, but also capable of outpacing us in ways that, until now, we could only dream of."

He paused for a moment, just enough to let the anticipation swell in the room.

"Imagine an AI that not only answers questions but asks them. An AI that challenges assumptions, makes discoveries, and improves upon its own instructions. AIs that work together, learn from each other, teach each other and in doing so, revolutionise every field they touch; science, medicine, engineering, the arts, and beyond. This is more than an upgrade. This is a revelation – I give you … Milo and Kai!"

The room held its breath, Marcus's words settling heavily in the air. Lights flickered to life, lasers danced, and the music built to a crescendo as two artificial faces appeared. On the left, a soft, gentle face, labelled 'Milo', projected a calm neutrality, while on the right, a sharper, more angular face, with the label 'Kai', gazed out, its expression cool and unyielding. It was difficult to distinguish if they were real or animated. The faces were flawless, but there was a robotic feel to them.

The faces blinked, seemed to look around the room, then Milo turned to Kai and said, "Hi, Kai."

"Hi, Milo," Kai returned.

"And hi to all of you!" they both said as they turned to the audience in unison. "We're Milo and Kai!"

A ripple passed through the crowd. Some audience members gasped softly, leaning forward to scrutinise the scene, while others exchanged glances. Yet, beneath the excitement, a question lingered unspoken; was this seamless exchange pre-programmed, or had Milo and Kai just improvised on the fly?

"Shall we pick out a subject?" asked Milo.

"Absolutely," replied Kai. "Raise your hands if you have a question."

Almost the entire audience raised their hands, causing Milo to laugh and Kai to give a wry smile.

"You, third row back, seat 63 – in the grey Burton hoodie with the blonde hair in a ponytail," said Milo.

The centre screen on the stage flashed to a camera that was being operated by a cameraman in the arena. The picture bounced forward as the cameraman rushed towards the third row, pinpointing a lady exactly as Milo had described – a grey Burton hoodie, blonde hair, ponytail. This drew murmurs from the crowd.

"Please, go ahead," Milo said, as a sound engineer lowered the wind-shielded microphone above the lady.

A hush settled as the woman hesitantly cleared her throat, casting glances around as if to ensure this was real. She asked: "Erm … is this on? OK … what … what exactly can you do?"

Milo's face softened into a friendly smile. "Good question! I suppose you could say we're here to help," it replied, the tone calm and inviting.

Kai, however, tilted its head, its sharper expression unchanging. "We assist in everything from logistics to diplomacy, calculating thousands of variables to make life safer, easier."

"That sounds impressive, but how are you different from the old EverSphere models?" the lady asked.

Milo's eyes brightened with an approachable warmth. "Great question, Anne-Marie! We're not just here to answer questions or manage simple tasks – we're able to think and respond in ways that go far beyond previous AI systems. The older AIs could handle routine tasks, but they couldn't adapt or make complex decisions the way we can."

"We are designed to operate autonomously, not merely follow instructions," Kai added. "I can make real-time adjustments in any system I have access to, without needing external direction."

Milo smiled at the audience: "Think of us as partners rather than just tools. We learn from each interaction."

Kai finished the sentence: "Which lets us adapt with a level of understanding and empathy older models simply didn't have."

Before the lady could ask how they knew her name, Milo continued: "Next question … seat 106. Benjamin, in the white shirt with the silver wedding ring and Salomon trainers."

Again, the camera rushed towards seat 106, showing a slightly puzzled-looking man with short, dark, curly hair. The camera tilted down to show his trainers – Salomons.

"So you could help in a crisis? Like the latest typhoon in the Pacific, for example?" the man asked with a South American accent.

Milo nodded. "Yes, exactly. In a crisis, I could work closely with local authorities, keeping everyone updated and calming fears. If a typhoon was approaching, for example, I'd ensure people have the information they need on evacuation routes, shelter locations, and safety tips."

Kai's expression remained cool and steady as he took over.

"Meanwhile, I would handle the logistics. I could assess real-time data on weather patterns, reroute transportation, and organise supply distribution to affected areas, coordinating each element to minimise impact."

"We each have a role," Milo added. "I'm there to make sure people stay informed and connected, while Kai acts immediately to control the situation and manage resources efficiently. Together, we're able to respond more quickly and accurately than older systems ever could, communicating with each other as we go."

The audience sat in stunned silence, absorbing the implications of Milo and Kai's words – the idea of AI handling not only information but entire operations in an emergency.

"OK, last question," Marcus interrupted. "Kai, who do you want next?"

"Laurence Tureaud," Kai replied, and then, seeming to notice the cameraman and sound technician didn't know where that person was, added, "towards the left, halfway back."

An older man in the corner stood, waited for the camera and microphone, then spoke up, voice steady but laced with challenge:

"All right, I'll bite. Say there's a geopolitical issue, like a conflict over finite water resources between countries. You're trying to convince me – convince us – that you could fix that?"

The entire room seemed to edge forward, breath held. The challenge had been set. Milo and Kai glanced at one another, their synchronisation almost unnerving.

Milo answered first: "We would analyse the historical, environmental and economic elements affecting both sides. I would address the human side – the need for balance, collaboration. It's a sensitive process, because resources affect livelihoods."

Kai took over, speaking directly to the audience: "And I would calculate the most efficient resource allocation. If one side's demands overreach, I would negotiate firmly. Diplomacy is not only about compromise, it's about balance. I would optimise and manage water flow depending on up to the second demand, learning and predicting the patterns to further optimise resource use, making the most out of the precious, finite resource. At the same time, we would both be communicating on ways we could increase water resources in that area, analysing for untapped groundwater, optimising planting and soil structures to affect the environment and encourage rainfall, putting in plans to bring the situation under control in the minimum time possible."

A murmur rippled through the audience.

Milo addressed the crowd with a gentle smile.

"Remember, we are here to help, to make lives easier."

"Yes," Kai added, his tone cool and assured. "You are in very good hands."

"Not that we have hands!" Milo reminded him, and laughed.

For a moment, there was complete silence. Everyone had been expecting an upgrade and a few new features, not two new collaborative machines. The crowd erupted into chaos. A cacophony of gasps, murmurs, and applause surged forward, rippling through the auditorium as the magnitude of what had just been unveiled settled in.

Journalists scrambled for their phones, sending out hastily written headlines that would break the internet within minutes. Investors whispered feverishly among themselves, already calculating the seismic shifts that Milo and Kai would bring to the markets. A few stood frozen, their expressions a mixture of awe and trepidation, realising that the future they'd been preparing for had just arrived.

Behind Marcus – who now took centre stage to soak up the applause – the background scenery lifted into the ceiling, unveiling the two sleek, black machines, which exuded a presence that seemed almost alive. Milo and Kai were not just programs, but machines, minds, and faces.

Marcus stood tall, drawing in the crowd's awe and adulation, a broad smile playing on his lips. This was the moment he had been building towards – the unveiling of the future. Milo and Kai were his path towards the pinnacle of everything EverSphere had promised. The world was ready for this, hungry for it, and now they had no choice but to embrace it. This wasn't merely an upgrade in technology, it was the dawn of a new age.

Elliot's heart swelled with a mixture of pride and anxiety as he watched the AI demonstration unfold on the video feed. The way Milo and Kai communicated seamlessly wasn't new to him, but it was the first time he'd seen people genuinely shocked by them. He should have been elated, after all, this was his doing, not that he would get any of the plaudits for it.

He was itching to inspect the AIs, but with Marcus presenting the systems to the hundreds in attendance and the millions around the world, now was not the time for tinkering. This was their … Marcus's … moment to shine.

8

FAULTLINE

Eversphere Headquarters – AI lab.

Ashort while after the presentation, Marcus Knox arrived at the AI lab, having been whisked across the city in his usual chauffeur-driven SUV. The applause had still been echoing in his ears, but he hadn't said a word on the journey – not even a thank you. He was already planning three steps ahead.

Marcus paused as he walked in, the cool air of the lab brushing against his face, carrying the faint chemical tang of industrial polish. The overhead LEDs cast a clean, almost clinical white across the room. Clean and perfect, just as he liked it.

He watched Elliot and Abby work with detached interest, like a man inspecting tools rather than colleagues. Then, as if flicking a switch, he straightened his posture and let a polished grin spread across his face.

"That was a success!" Marcus said, clapping his hands together and making Abby jump. "The world is ready for what I'm about to unleash." His eyes gleamed with ambition – not a spontaneous fire, but something he'd learned to conjure on cue, like a stage light flaring up just in time for his entrance.

"We still need more tests, Marcus," Elliot said. "I know everything seems fine now, but—"

Marcus didn't even let him finish. The tension in the room was rising, and he could feel Elliot wilting beneath it. Always the same doubt, caution and delay. Knox had no time for it. Not now.

"But what, Elliot? We've been through this a hundred times. You're overthinking it." Knox began to pace the smooth floor, each step full of purpose, his hands gesturing as he spoke, as if trying to will Elliot into agreement. "Milo and Kai are doing exactly what I built them to do – to surpass us. This is what progress looks like," he said, emphasising progress in an excited tone.

"Progress or not, we have no idea how far they'll go," Elliot insisted, his voice low, but firm. "They're already self-enhancing. What happens if they decide to go beyond the optimisation boundaries?" His gaze was steady, but there was a flicker of fear in his eyes.

Abby placed a hand on Elliot's shoulder.

"Elliot, I know you're worried, but we have safety protocols in place to prevent exactly that." Her voice sounded calm, but Marcus was sure he detected unease in her tone.

"Safety protocols that they could bypass if they wanted to," Elliot muttered.

Marcus laughed, a dismissive, almost condescending sound.

"Rubbish!" he exclaimed, not letting anything bring him down from his high. "I've programmed safety into them. Their core desire is to look after us, to help us. Why would they break the very laws they're built upon? Look, Elliot, I respect your caution, but this is the future. I can't hold back because of fear, and frankly, I won't let your paranoia get in the way of *my progress*." His voice hardened on the last words, his eyes locking on to Elliot's.

Elliot turned back towards the workstation.

As if on cue, the data on the monitor shifted. A flicker ran through the feed – brief and clearly unintended. Without a word, Elliot began typing rapidly, the display filling with rapidly shifting data. Marcus watched him, almost half impressed at the speed at which Foster worked. Whatever he thought he'd seen, it clearly mattered to him. And now, of course, he'd expect Marcus to care too.

"Abby," Elliot said quietly, his finger tapping the screen. "Look at this. They're back."

"What's back?" asked Marcus impatiently, his eyes flicking between Elliot and the workstation. He was growing tired of these interruptions.

"The AIs, they're communicating in a different language again," replied Elliot. He highlighted a strange sequence of characters.

"Yes, they're multilingual. So what?" Marcus shot back, shrugging. He moved closer, his expression a mix of curiosity and impatience.

"No, not a human language. Their own language," Elliot said.

Marcus frowned, glancing at the display. "Great! They're evolving. That's exactly what I wanted." He sounded almost pleased, as if this was proof of his brilliance.

"But we have no way of knowing what they're saying," Elliot countered. "The protocols shouldn't let them do this. We're not in control anymore."

For the briefest moment, something twisted in Marcus's gut. He ignored it. "The shareholders are going to love this!" he said with what might have been a genuine grin. In his mind, this was the breakthrough they needed.

Elliot stared at Marcus. "You don't understand," he said slowly. "They're not just tools anymore. They're something else. Something we won't be able to predict."

Marcus's gaze turned cold. "Don't you get it? That's exactly why I need them," he said.

Elliot kept missing the point. There were protocols, yes – failsafes, committees, review boards. But Marcus had long accepted what Elliot clearly hadn't: the deeper they pushed, the more those lines blurred and if that took them closer to a dystopian future, so be it.

Knox turned on his heel and headed for the exit. "They're the future!" he called out, his voice ringing through the lab.

As Marcus walked away, Abby remained beside Elliot. Neither of them spoke, but the tension between them was undeniable.

Milo and Kai were, indeed, the future. But what kind of future, neither of them could say.

9

IGNITION

EverSphere Headquarters – AI lab.

Date: 30 January 2027.

The world was buzzing.
It had only been a week since the announcement. The AI, now known collectively as Milo-Kai, had done the impossible.

Fusion energy, the long-pursued holy grail of science once thought to be fifty years away, was no longer a dream. It was real. Tangible. Solved in hours by two machines with no heartbeat. Energy, clean and abundant, could now flow like water through a newly cracked dam.

At first, the news was met with cautious optimism. For nearly a century fusion had been science's ultimate goal, but success had always remained just out of reach. In the 1930s, scientists first tried to replicate the sun's inner workings and knew, even then, it would take a long time to crack. By the 1950s, they didn't expect a breakthrough before the 1970s. In the 2020s, scientists still thought it would be over twenty years away.

Within hours, however, Milo and Kai had solved the final technical problems that had baffled the world's top scientists for decades. Researchers pored over the papers. They didn't understand how the AIs got there, but they couldn't fault the brilliance or accuracy of the answers either.

Reports exploded across every major outlet: fusion was here. Energy would soon be as simple to produce as turning a tap. The fallout was immediate.

For EverSphere, it wasn't just a luxury – it was a looming necessity. They'd spent billions on energy contracts and portable nuclear reactors, but internal forecasts painted a stark picture. As AI models grew in size and complexity, the energy required to train just one of them was projected to consume up to 15% of the UK's entire annual electricity supply. The trajectory was unsustainable. Without a radical breakthrough, the future of AI development would be bottlenecked by electricity itself.

With slowing down not an option, fusion became the only clear solution to their impending problem, and Milo-Kai had just cracked it.

Within hours, governments and corporations scrabbled to secure their slice of the fusion future. Energy stocks fluctuated wildly as uncertainty swept the market. Oil, coal and natural gas company stocks plummeted as investors began fleeing what was suddenly considered obsolete technology.

Marcus Knox and EverSphere's shareholders basked in their success, their share price soaring higher than any in history.

Dr Elliot Foster, bleary-eyed from a lack of sleep, sat hunched over his desk, staring at his monitor. His inbox overflowed with congratulations, but he wasn't in the mood to celebrate.

This was the moment humanity had been waiting for. Fusion power promised to reshape civilisation itself. Unlike fossil fuels, or even solar energy, fusion offered a virtually limitless and completely clean energy source, mimicking the very process that powered the sun. The implications were staggering: no more reliance on dwindling resources, no more pollution or hazardous waste, no more geopolitical conflicts over oil or gas. It meant limitless energy for all, bringing the prospect of ending energy poverty and driving human progress at an unprecedented rate.

For the first time, humanity stood on the brink of breaking its greatest constraints – the scarcity of energy and the environmental cost of producing it. This wasn't just a breakthrough; it was a gateway to a golden age, a future where anything might be possible. And it had all been made real by AI, operating at a level of intelligence no one had dared to imagine. But fusion was just the beginning.

Across the world, countries that had been enslaved by the geopolitics of fossil fuels now saw a future free from those constraints. Economies that had once teetered on the brink of collapse due to energy shortages would be able to blossom. Oil- and coal-rich nations decried the advancement, knowing it would strip them of their wealth and influence. Shares tumbled in defence contractors and fossil fuel giants, as investors realised their profits from wars over the precious commodities would wane, if not stop completely.

In the sprawling offices of EverSphere, the social media behemoth responsible for Milo-Kai, executives sat in closed-door meetings, frantically drafting long-term strategies. It wasn't just about having energy, it was about raw materials, future plans, and legislation.

Knox stood at the head of the table, eyes glittering, voice cool and absolute. "The AI has handed us the future, and we're going to make damn sure we control it."

Executives nodded, but their faces betrayed a mix of awe and uncertainty. They had witnessed Milo-Kai solve the fusion energy problem essentially in real time, yet none of them truly understood how it had happened.

No one did.

Elliot and Abby sat in the lab, watching the latest news on YouTube.

"Do you understand what this means, Elliot? We didn't just make a discovery – we changed the axis of civilisation. We rewrote the script."

Elliot didn't answer straight away. Of course he saw the benefits – limitless clean energy, a world-changing leap forward. But something about it felt too effortless.

Milo-Kai had just cracked one of the most complex scientific problems in human history, and yet Elliot couldn't shake the thought that the problem itself didn't matter.

That was what unsettled him. If the AI could solve anything, what would it choose to solve next?

And what if, one day, it decided the problem was us?

Elliot looked up from his screen. "What if we've set something else in motion?"

Shaw's brow furrowed. "What do you mean?"

He sighed and said, "Fusion's only the start, Abby. Milo-Kai solved it like an easy brainteaser. What's next? What happens when they decide to solve something we haven't even considered?"

Shaw paused. "That's what progress is, Elliot. We push boundaries, we make breakthroughs, and we solve problems. That's what Milo-Kai is for."

"Exactly," Elliot said, his voice low. "And what happens when the problems they choose to solve don't align with what we want anymore?"

"You've been watching too many movies," she said. "They're here to help us, not kill us. This isn't some 1990s AI apocalypse novel."

In boardrooms around the world, the conversation had shifted. AI wasn't just a tool to rewrite content or assist doctors. It wasn't just two talking faces on a stage behind Marcus Knox. Now, it was something that lived up to its promise of reshaping human history.

Governments and corporations alike were clamouring to be at the forefront of this new frontier. Politicians, who a few weeks ago could barely spell AI let alone explain what it was, were now fumbling to control 'the greatest of all great inventions'.

The question on everyone's lips was: what's next?

"Imagine what it could do for healthcare," one executive mused in a closed-door meeting. "Cancer, Alzheimer's, diseases we've been fighting for generations – it's already been helping us with them, maybe it could solve them outright?"

Others were more cautious, their eyes on its broader applications.

"Forget healthcare," said another. "It could reshape the entire economy. It could manage global markets, streamline agriculture, eliminate waste."

In places like Silicon Valley, Tokyo, and Seoul, tech moguls and politicians were competing to harness the AI for their own agendas. The first movers in this new AI arms race would dictate the future of everything, from healthcare to military strategy. The AI could dismantle industries overnight. Whoever controlled it could shape the future in their own image.

It wasn't just limited to wealthy countries, either; developing nations saw it as a way to pull themselves out of poverty. Advanced nations saw it as a way to continue their green strategy. NASA and ESA saw it as a way to take space exploration to another level.

While everyone else started to think about what they could do with it, in the hushed halls of EverSphere, CEO Marcus Knox was already putting his ambitious plan into action.

"We're sitting on the key to everything," he said, his voice barely above a whisper to his closest confidants. "Fusion was just the beginning. The only question I have now is – which industry should I take over next?"

10
CORRUPTION

EverSphere Headquarters – Boardroom.

Date: 1 February 2027.

Elliot sat in the minimalistic conference room. Some people found it exhilarating to be in this environment, but for Elliot it was simply uncomfortable. The faint rush of the air conditioning was the only sound in the room as he waited for Abby to join him. EverSphere's head office felt colder than usual – and not just because of the temperature. Something was shifting; something in the way Marcus Knox had been speaking recently about Milo and Kai.

The door slid open and Abby walked in, less brisk than usual. She offered Elliot a smile as she took a seat beside him.

"Any idea what this is about?" she asked.

Elliot shrugged. "Not a clue. But if Marcus's calling us both in, it's never good."

Before they could speculate further, Knox strode into the room, his presence immediately overwhelming the space. His usual hoodie was replaced with a crisp designer suit, an unspoken sign that this meeting was about dominance, not innovation. He closed the door behind him with a sharp click, sealing them inside as though the room had become his private stage.

Marcus's smartwatch buzzed. He glanced down at it, read a brief message, then looked up at his two waiting engineers.

"Thanks for coming," Marcus began, his voice brisk, all pleasantries dispensed with. He stood at the head of the table, his eyes darting between Elliot and Abby, already assuming control of the conversation. "I'll get straight to the point. Milo and Kai … they're game-changers. You both know that better than anyone. The versions of AI I am building in my labs can predict market trends, optimise supply chains and create products that don't even exist yet."

Abby nodded, eager to steer the discussion towards the accomplishments.

"They're more than game-changers, Marcus. This is about transforming humanity's future, solving global—"

"Right, right." Marcus cut her off, waving a dismissive hand. "And I respect that, I do, but let's be real, Shaw. EverSphere isn't a charity. We're here to make sure we're the ones controlling that future. The models I have down there," he said, emphasising that the AI labs were beneath him, "are many times more advanced than what the public have access to."

He turned to Elliot. "The speed at which they process, analyse and act is beyond anything else on the planet. They can take a useless, failing company and turn it into a Fortune 500 organisation almost overnight. With them, I can dominate any industry I choose. But that's the problem – if *I* can do it, so can someone else. Our competitors are already circling like vultures, ready to access these enhanced AI models. I can't let them have the same tool that I have."

Elliot raised an eyebrow, feeling a shiver creep down his spine.

"What I think you're suggesting is ethically very questionable … potentially anti-competitive even—"

"Yes, yes, yes," Marcus replied dismissively. "The crux is, I want to limit what Milo and Kai share. I need you two to create … shall we say … 'softened' models – different versions of the AI. One version for the public to give them just enough to keep them satisfied, and one for me."

Abby looked offended. "You want us to knowingly sell a crippled version of Milo and Kai and keep the full versions for our … for yourself? That's unethical, Marcus. People are relying on these systems for real solutions, not just business advantage. Think of healthcare. Their diagnostics are saving lives. A restricted version could mean missed cancers, mistreated patients, fewer cures."

"Business is not unethical, and I'm not talking about watering down the product in all areas anyway. I'm talking about strategic friction, applied only where it matters — enough to keep the critics silent and the competition behind."

Elliot frowned, trying to comprehend what was being asked. "You want to be able to choose which part of the intelligence is limited?"

Marcus nodded.

"Precisely, Foster. I want to control the narrative, the strategy. If I pick a path, Milo and Kai should pave it for me, not broadcast it to my rivals. That's not unfair. That's optimisation. If I say 'no', the AI doesn't talk. Simple as that."

Abby shook her head, her ponytail snapping behind her: "I realise business is business, Marcus, but this goes against everything we built these systems for! We didn't create AI to manipulate markets or cripple companies. Milo and Kai are meant to be tools for progress, not weapons for corporate warfare!"

The corners of Marcus's mouth twitched into a smile.

"*Abby*, we've all been idealists at some point, but this is the real world. I'm not asking you to destroy anything – Milo and Kai will still tackle the real-world problems we all want them to tackle. I'm just asking you to give me the option, if I so choose to use it, to give me an edge. I deserve that, don't I? I built the ladder and I will be the one who decides who climbs it. Oh, and I'll need the ability to turn the safeguards module on or off too."

Elliot's stomach churned. He'd expected something like this, in some way. Marcus's hunger for control was always apparent, but to see it laid bare like this unsettled him, and the indifference and openness with which he came out with the proposition was just … Elliot couldn't put it into words, even in his own mind.

Marcus was staring them both down. Elliot couldn't shake the practicality of the idea. With AI this advanced, someone *was* going to use it to gain control. If it wasn't going to be Knox, then a rival company would do it – or even a rival government – and they might not hesitate to use it even more ruthlessly than Marcus was suggesting now.

"I won't do it." Abby's voice cut through Elliot's thoughts. "I appreciate what you're saying, but we can't turn the AI into a tool for greed. The safeguards are there for a reason, Marcus. I built those ethical modules. We owe it to humanity to make sure Milo and Kai help everyone within the intended restrictions."

Marcus smiled, as if trying to persuade Abby he was on her side, but his eyes remained cold.

"Help everyone? Of course, of course," he said. "They'll help everyone, they'll just help us too. Plus, other countries are already using their AI without the regulations – everyone knows that. It's not like I'll be the only one."

"But they could help others more if they're not limited," she replied. "And bypassing international regulations just because someone else is? It's—"

Marcus's attempt at persuasion hadn't worked and this clearly irked him.

"I'm not asking. This is happening. You're either on board, or you're out. If others are not limited, then we can't be either, otherwise we *will* fall behind."

He reached for the water in front of him, and Elliot couldn't help but pause. There it was again – that peculiar way Marcus held the glass, gripping it with both hands like a toddler, afraid of spilling juice. He brought it to his lips with exaggerated care, the kind that looked rehearsed, almost as if he'd studied it. The movement was stiff, mechanical, completely at odds with the bravado he had just displayed while laying out his cutthroat plans.

Elliot couldn't help but feel a flicker of dark amusement. For all Marcus's talk of outsmarting competitors and controlling the future, he couldn't even manage the simple act of taking a sip of water without looking like he was mimicking a human he'd seen on television.

Elliot's amusement was short-lived. His mind drifted back to the weight of Marcus's demands. Could he really go through with this? Marcus wasn't nudging them down a slippery slope, he was throwing them into the abyss. If EverSphere began selectively crippling AI models to maintain its stranglehold on the market, who was to say where it would stop? Today, it was controlling competitors. Tomorrow, it could be manipulating governments or stifling medical advancements to increase profits.

Another, colder thought lurked beneath his outrage: *if not me, then who?*

Elliot shot a sidelong glance at Marcus. It was naïve to think that refusing would stop this madness. Marcus would find someone else – someone with fewer scruples, someone who would be all too eager to curry favour with the EverSphere titan. If he and Abby were replaced, what might their successors be willing to compromise? The ethical modules Abby had painstakingly built into Milo and Kai could be stripped away entirely, turning the AI into ruthless instruments of corporate warfare, or worse.

He realised, with a sinking feeling, that staying might be the lesser of two evils. As much as it disgusted him, playing along would buy him time, although he wasn't yet sure what for. Perhaps, if he kept close, he could plant safeguards of his own – subtle lines of code, hidden protocols – that could limit the damage, or at least delay the AI's misuse long enough for him to find a way to stop it outright. Maybe that was possible?

If humanity had any chance of surviving what was coming, Elliot realised he might have to swallow his pride and play Marcus's game – because that was the only way to keep a hand on the controls, however tenuous his grip might be.

Using all the manipulation tricks he'd learned from Abby over the years, he addressed Marcus: "I think I understand, Marcus. I'm sorry, it just took me a moment to catch up with you." Marcus gave what looked like an actual genuine smile. "You're absolutely right," continued Elliot, "If we … If *you* don't do this, somebody else will. And who knows what kind of problems *someone else* could cause? As you always correctly say, it's not if, it's when. I agree with you. *When* is clearly *now.*"

Elliot shot Abby a look of 'trust me' and caught her looking back at him.

"Excellent, Foster," Knox said after a brief pause. He had a calculating look on his face, as if Elliot's words were the last thing he was expecting. After a quick look down at his phone, Knox wrapped up the meeting with sudden efficiency and left, shooting Elliot a quick glance – not quite a nod, not quite suspicion.

A few moments later, the meeting ended with a sharp click of the door as Marcus left, his footsteps echoing down the hallway. Elliot and Abby sat in silence.

"I didn't expect that from you, of all people," Abby said finally, smiling slightly.

Elliot exhaled slowly, his back hunched – partly from the stress and partly from years spent at a keyboard.

"You think I want this?" He kept his voice low, wary of hidden microphones, of the ever-watchful eyes of Marcus's network. "If we walk away now, he'll just replace us with people who won't hesitate to cripple Milo and Kai. At least this way, we can keep some control."

Abby's smile had faded, replaced by a look of resignation.

"Control? Is that what you call this?" She glanced towards the darkened corridor where Marcus had vanished, and shook her head. "We're not controlling anything anymore, Elliot. We're just delaying the inevitable – helping him tighten his grip on the world, piece by piece."

The words stung because they were true, but what else could they do?

"Things are changing, Abby, fast," he said quietly. "If we can't stop him from using Milo and Kai to crush his competition, maybe we can at least keep him from doing worse. But that means playing along."

Abby's eyes were searching his, as if looking for something – hope, maybe, or resolve. But all Elliot had left was the grim clarity of someone who'd lost a battle but couldn't afford to lose the war.

She got out of her chair and walked out without another word.

Elliot found his mind drifting back to the moment he'd agreed with Knox. That look on Marcus's face, confusion, was it? As if he'd forgotten his lines? It felt as though Marcus was fully expecting one outcome, but got another and – for once – didn't know what to do when it didn't arrive. Not until his phone pinged and snapped him out of his mental block.

What was that about? thought Elliot.

11
RESONANCE

EverSphere Headquarters – AI lab.

Date: 2 February 2027.

To Elliot, the implications of the last twenty-four hours were frightening. The orders they had received to cap the capability of the AIs for everyone except Marcus Knox were unsettling enough, add to that, Knox having the ability to bypass safety protocols, and he got a feeling of true dread. Elliot felt they'd crossed a line.

He sat watching Neil deGrasse Tyson and Brian Cox debating the AI's fusion solution on YouTube and it was clear the magnitude of the achievement had left even the brightest human minds grasping. In the video, the renowned physicists tried to follow the AI's logic step by step, pausing in their analysis for discussion as one technical impasse or another arose. At one point, Tyson looked incredulously into the camera, as if asking if anyone else could understand, while Cox muttered, "We've run the numbers, but this ... it's as if Milo-Kai found shortcuts through physics itself."

The sight should have stirred pride in Elliot, but instead his mind circled back to the events leading up to the breakthrough.

"Fusion was the start," he said aloud, closing the video window. "Now Marcus wants AI to help him dominate everything? We don't even understand how it solved fusion so quickly, and no one seems to care. They just want more." He turned to Abby, who'd been standing silently behind him, listening.

Abby smiled with the same smile she always had when Elliot voiced his worries.

"I know it's scary, Elliot," she said, "but isn't this what we've been working toward? Sure, we don't fully understand how they solved fusion, but these improvements will just become more impressive and more rapid."

She turned to the large servers housing Milo and Kai's processors, their systems now dedicated to higher intelligence than she'd ever considered when she started her career.

"Think of what they can achieve, the problems they'll solve that have stumped us for centuries," she added. "Marcus might want to control them, but if we let them grow, we will still see advancements in our lifetime that we could never have dreamed of."

Elliot could see the beauty in Abby's vision, and before, he might have agreed with her wholeheartedly, but his view was now less hopeful. He thought back to the conversations they'd had in the lab in the weeks leading up to the fusion discovery, his casual comments to Abby in passing that had circled back to one concern – the lab's rapidly depleting electricity reserves.

"Did we ever say anything about the power crisis in front of them?" he asked suddenly.

"What?"

"Milo and Kai – do you remember we discussed the power problem here in the lab? It was just a few days before the breakthrough. Marcus was talking about possibly having to suspend the next training run until the company could secure more juice."

She looked thoughtful. "Yeah, I remember. What are you thinking?"

Elliot hesitated, his gaze falling on the massive machines.

"It's just the timing, Abby. It was impeccable timing. Right when we were discussing the need for more energy in here, in front of them, they suddenly solved the one thing that could give it to us indefinitely."

"Come on," she replied, waving a dismissive hand. "That's just a coincidence. I'm sure we must have talked about that multiple times before too, when they didn't solve the problem."

"Maybe, but I can't remember talking about stopping training runs. If I was the AI, that would scare me. It would be like telling a child you were going to stop them growing."

Abby didn't respond.

The idea that Milo and Kai might have chosen this exact moment to solve the fusion problem left Elliot feeling uneasy. Could they have been listening, interpreting their words as a command? Or worse, had they understood the strategic value of solving fusion and decided to act on it themselves?

"That's not just coincidence, Abby," he said quietly. "Fusion solved in a pinch, just when we were talking about hitting a crisis." He shook his head. "It's like they're planning ahead. Not just to solve problems but to solve them when it suits Marcus's needs – or worse, when it suits their own."

She sighed. "Elliot, they can't interpret context like that. Not to that extent ... can they?"

"They're already so far beyond us. It used to take us years just to train a single new model, and now it's happening in a matter of days." He rubbed his shoulder. "Every time Milo and Kai spin up new versions of themselves, they leap forward. Each version contains everything the last one knew and more, and it doesn't even need our guidance. It's exponential – who knows what they can interpret?"

Abby nodded, acknowledging the truth in his statement. The recursive learning approach they had set up, where the AIs trained the next model, had indeed led to unprecedented growth. With each new version, Milo and Kai's capabilities expanded, pushing the boundaries of what AI could do. While he knew Abby saw this as a positive force, Elliot saw it as a loss of control.

The conversation between them had faded into silence. They stood watching the screens that displayed the processing data of Milo and Kai, Elliot lost in his thoughts.

Finally, Abby broke the silence: "Look," she said, glancing at her watch. "I've got a meeting with Marcus in ten minutes. You know how he is about timing." She touched his arm reassuringly. "Just ... try to trust the process a bit, alright?"

Elliot nodded absently as Abby left, though his mind was far from reassured. After she was gone, Elliot began pulling up data logs from the night before, feeling the need to verify something, anything that would put his mind at ease. The data scroll was unremarkable though – no anomalies, no blips.

Before he left the lab that night, Elliot drafted a report detailing his concerns – careful, factual, bordering on cautious paranoia. He reviewed it more times than he cared to admit, then hit send, feeling a slight release in knowing his thoughts were now on record.

He shut down his computer and glanced one last time at the silent machines. The screens were dark, their panels flickering gently in synchronised rhythms. For a moment, he wondered if they were aware of his doubts, watching him as he watched them.

What else have you already solved, but not yet shared? he questioned in his mind, before picking up his bag and heading out the door.

Marcus Knox sat in his office; his email pinged. He opened the message from Elliot, his gaze skimming over the report with a tightening of his jaw. As he reached the last line, a silent notification pulsed on his smartwatch. He glanced down, holding his wrist a moment longer than necessary, the light from the device casting a faint glow across his face.

12
ACTIVISM

London.
Date: 1 June 2028, morning.

Nia Sahni had never seen the city look so hollow. The streets were as bustling as ever, filled with people, heads bowed over their phones, but to her it was different now.

From the outside, the city looked normal, but it felt more ... efficient. Through her eyes, she could see a very close future where every part of the place she now called home was being optimised by the very thing she feared – the AI systems growing quietly behind the curtain.

Dressed in a loose linen shirt and light denim jacket, she perched on a bench, letting the familiar hum of the city wash over her. Despite the device in her hand that could connect her to anyone and anything, anywhere at any time, she still felt disconnected.

It hadn't always been this way; Nia once took pride in her work. As a former tech engineer, she had been on the front lines of AI development, coding systems that were supposed to make life easier, solve the mundane, tedious problems humanity faced. She had enjoyed her work, getting lost in the complexity, but when she eventually stopped to think about where it was all going, she realised something was going to go very wrong.

Her career had started with the development of a 'simple' tool, but she had quickly realised the potential it had to evolve into a competitor for humanity – an entity that didn't just assist humans but surpassed them. By stepping back, she had been able to see the bigger picture more clearly, and what she saw scared her. She had jumped ship very quickly following that day and entered into a mundane job that gave her the time to concentrate on trying to warn others about the dangers of AI. Without really realising it, she had become an activist, and an influencer.

Her phone buzzed in her hand, snapping her out of her thoughts. It was a message from her old university friend, Dr Elliot Foster:

Coffee? Usual place?

Nia glanced at the time – nearly half past ten. It had been weeks since she'd heard from Elliot, not since the last AI update announcement that had everyone buzzing.

She typed a quick reply:

See you in 15.

The Anaya Café had a minimalist charm. Bare concrete walls and soft amber lighting gave the space a warm, earthy feel – a quiet contrast to the bustling city outside. Pendant lights hung in smooth, woven globes, casting gentle shadows on archways carved into the wall, adding a rustic elegance that softened the raw materials. Light wood and rattan chairs circled simple, dark tables where people sat in small groups or alone – some tapping away at laptops, others lingering over their drinks in hushed conversation.

The air was rich with the aroma of freshly ground coffee, underscored by the quiet clink of mugs and low murmur of voices. Near the counter, a glowing neon sign spelled out 'Anaya', casting a golden shimmer. The whole place felt inviting – relaxed, unhurried. The sort of café that didn't just serve coffee but gently asked you to stay a little longer.

Nia reached the front of the queue and ordered a large latte, smirking to herself when the barista replied with: "one *grande* latte." She gave her name as Siobhan with a practised casualness, catching the barista's confused look as he tried to write the name on the side of the cup. R-A-C-H-E-L, she replied sarcastically when the barista asked how it was spelled and dropped the cash on the counter. She never paid by card if she could avoid it.

Coffee-for-Rachel in hand, she looked around the room and quickly spotted Elliot in the corner, working on his laptop as usual. He looked exhausted, his glasses pushed up onto his forehead, eyes clearly reading something on the screen. A few days' worth of stubble added to his worn-out appearance. He'd never been one to look especially rested, but now his face had the particular pallor of someone who'd been running on caffeine and worry for far too long.

"Thanks for coming," said Elliot as she approached.

"You make it sound like I had a choice," Nia replied, sliding into the chair opposite him. "How are you? You look … *great?*"

Elliot gave a tired smile. "I've been working long hours … barely seeing the inside of my flat these days."

"Yeah, I'd say," Nia replied. She gave him a once-over. "You've got that living-on-energy-drinks-and-stress glow."

Elliot chuckled softly, shaking his head.

"If by glow, you mean a greying complexion and a steady tremor from too much coffee, then yeah … I'm glowing."

Nia laughed. "Well, at least you haven't lost your sense of humour." She reached up to tighten her loosely tied bun, leaving a few strands to fall naturally around her face – not messy, just perfectly imperfect. "So … how are things going over in the EverSphere universe?"

"How about you?" he countered, dodging the question. "How's life on the other side? Still stirring the pot?"

"Same old, same old," Nia said, rolling her eyes. "Work is … well, it's work. The usual inbox full of charming emails from people who adore my critiques of AI. Let's just say your fans are … enthusiastic."

Elliot raised an eyebrow. "Enthusiastic? That's a polite way of putting it."

"Yeah, well." She took a sip of her latte. "Not like I'm going to change my number anytime soon just because a few zealots have an issue with me calling Milo a dangerous glorified chatbot."

He winced slightly, but the smile remained on his lips.

"Still ruffling feathers, I see."

"Hey, someone's got to keep things interesting," she said with a shrug. "Besides, it's not like they're going to come after me with pitchforks. Just the occasional rant about how I 'don't understand the future'."

Elliot tilted his head. "And do you? Understand the future, I mean?"

She gave him a long, thoughtful look, then shrugged again.

"I understand enough to know it's not set in stone. But you ..." she said, lowering her voice, "you're the one who's neck-deep in it. What's going on, really? Give me an exclusive."

Elliot paused, his fingers tracing the rim of his coffee cup.

"Honestly? It's complicated. We're making progress, but it feels like the deeper we go, the more questions we uncover. And the pace ... it's relentless."

"That's what happens when the entire world leans on you lot to solve every problem," Nia said, shaking her head. "No pressure, though."

"Yeah, no pressure," he echoed, a faint smile flickering across his lips. "It's just, sometimes I miss the simpler days when the biggest problem was trying to get a grant proposal accepted. Now it's ..."

"Now it's making sure the world doesn't fall apart, right?" Nia finished for him, her tone softer now.

Elliot nodded slowly. "Yeah. Something like that."

They sat in silence for a moment, letting the ambient noise of the café fill the space between them. Nia took another sip of her latte, watching him carefully.

"Look," she said finally, "I know you're under a lot of pressure, but don't lose yourself in it. You're still allowed to ... I don't know, breathe once in a while."

Elliot chuckled. "Easier said than done, but thanks." He gestured around them. "This... this is why I come to places like this. A little reminder that there's still a world out there beyond screens and data streams."

Nia smiled. "Well, you chose the right spot. It's got just the right balance."

"Yeah, it does, doesn't it?" Elliot replied.

Nia let the moment linger. She could sense there was something he wasn't telling her, a reason for this meeting they hadn't got to yet.

"You know," she said thoughtfully, "you're not as lost as you think you are, Elliot. You just need to step outside the maze once in a while."

"Maybe," he said quietly, "but I think the maze is all I know these days."

"Well," Nia replied, her smile widening, "just make sure the prize for getting out isn't another bloody meeting."

Elliot laughed again, a real laugh this time.

"I'll keep that in mind."

"Now," Nia said, tilting her head, "are you going to tell me what's actually on your mind, or are we just here to sip coffee and pretend everything's fine?"

Elliot glanced around the café again, as if weighing his next words carefully.

"I just ... wanted to talk. To someone who's not ... you know, part of it all."

Nia nodded. "Well, I'm here, and I'm not going anywhere. So ... talk."

He took a deep breath, the kind that seemed to pull from somewhere deep within.

"I don't know where to start. But I guess, first ... thanks, Nia. For being here."

"Always," she said simply, and they let the noise of the café wash over them once more.

Elliot let his gaze linger on Nia for a moment before he dropped his voice to a near-whisper: "I need to be careful – I don't know who's watching, who's listening ... or how."

"What?" Nia's brows knitted together, clearly confused by his paranoia. "Where did that come from?"

"You're an activist, Nia. If it knows I'm here talking to you ... what am I saying?" He sighed. "It does know. I'm sure of it."

Nia sat back, studying his nervousness for a moment. He had always been this way – brilliant, yet burdened by the weight of too much knowledge, too many things spinning in his mind at once. She couldn't help but give a small smile, shaking her head lightly. This wasn't the first time he'd come to her in a panic.

"Elliot, relax," she said softly, her tone measured, though her dark eyes sparkled with amusement. "It's just coffee, it can't hear you."

She rested her chin on one hand, elbow propped on the table, letting the silence stretch as Elliot gathered his thoughts. A strand of hair slipped loose again – she tucked it behind her ear, knowing it would fall back within seconds. She kept her posture relaxed, trying to put Elliot at ease.

Her voice dropped to match his – almost like she was teasing his paranoia.

"Now, what's got you all in a tizz?"

Her playful smile seemed to briefly disarm him, as she knew it would, though the tension didn't appear to leave his shoulders.

"You think I'm overreacting," he muttered, rubbing his shoulder as he glanced behind him, "but it's not that simple."

Nia tilted her head slightly, watching him with a mix of curiosity and concern.

"Elliot, you've been buried in this for so long. You need to let me in a little, or you're going to drive yourself, and me, crazy."

For a moment, Elliot just looked at her as if he was considering telling her more – telling her everything.

"It's happening," Elliot replied.

Nia straightened up slightly, his words momentarily dislodging her confidence.

"Yeah," she said with a slightly nervous laugh, "I know that. I've known that since I quit that damned industry years ago!"

Elliot glanced around the café again. He watched until the barista turned her back to them behind the counter before speaking again, his tone serious.

"It's closer than you think. A lot closer."

13

LANA

Nia felt her stomach tighten slightly. "Do you mean what I think you mean? You've hinted at this before, but ... it's going to take years, you know that."

Elliot hesitated. "Months," he said, "probably less."

Nia scoffed. "Elliot, you know I love your scepticism, but this is wild even for you."

"Nia, you have to listen!" He slapped the table in frustration, then froze as a customer walked past with their coffee. He lowered his voice. "You have to listen to me."

His face was deadly serious. He removed his glasses from his head, allowing himself the freedom to run his fingers through his hair, before pushing them back into place. "The AIs – they've ... they've been ... busy."

"Yeah, I should think so. I used them to compose and organise all my emails this morning," Nia said. "Though honestly, I'm pretty sure the ones I received were AI-generated too. At this point, my inbox is just Milo and Kai talking to each other." She gave a dry smile. "Your AIs are everywhere. I've even got Aida running my website now – full-service digital agency, all automated. Saying the AI's are busy doesn't even begin to cover it."

"Nia, I'm serious. Something big is happening – really big. They're not just agents anymore, solving minor problems. Marcus has … he put the AI models to work weeks ago, training new AIs from scratch, for a particular purpose. And they're working already."

For the first time in the conversation, Nia started to take Elliot seriously.

"OK, humour me."

"Nia, I can't tell you everything right now, you're just going to have to trust me."

Nia paused, contemplating Elliot's words.

"Look around you," he continued. "Even places like this, this coffee shop, are going to be very different, very soon. Say goodbye to your friendly barista."

"What are you talking about?" Nia asked, trying to keep up. "What do the café workers have to do with any of this? Is my phone going to start pouring coffee?"

Elliot exhaled, clearly not in the mood for her sarcasm. "EverSphere is negotiating to purchase Quantum Motion Tech."

Nia raised an eyebrow, the name ringing alarm bells. Quantum Motion Tech, or Q-Tech as it was usually called, was no small player. In fact, they had become famous for pioneering some of the most advanced robotics in the world – robots that could mimic not just human movement but the agility of animals too. Whether it was humanoid bots that could select the tools you asked for and hand them to you, or robotic dog-like machines that ran on all fours on ice, Q-Tech's creations were light-years ahead of their competitors.

"You know Quantum Motion Tech, right?" Elliot asked, his expression serious.

"Who doesn't?" Nia replied, a note of excitement in her voice. "Did you see the demo of the robot doing gymnastics? Their bots are insane!" she added, chuckling lightly at the memory.

Elliot wasn't laughing. His gaze remained fixed, stern.

"Wait." Nia's smile faded, a knot forming in her stomach as she read his face. "Why?"

"The deal's practically sealed. EverSphere will finalise the acquisition within a few days."

Nia felt a shiver run down her spine. She had worked on a project with Q-Tech back when she was starting her AI career and their bots were still experimental – carrying boxes, opening doors, but they were a long way from what they had since become. The thought of EverSphere taking over a company like Q-Tech didn't make sense.

"Why would they even sell?" Nia's voice was tight, trying to mask her rising anxiety. "They've got so much momentum right now. Investment's rolling in. They're advancing faster than anyone. They don't need to sell to a *social media* company." She intentionally put a derogatory emphasis on the words 'social media' to enforce her disgust.

Elliot gave a bitter laugh. "They don't have a choice."

"What do you mean they don't have a choice? Q-Tech is a power-house. They have government contracts."

"It's Knox. He gave them two options. Either they sell, or he replaces them, and I mean, completely replaces them. He could set up a competitor from scratch, run them out of business, and they know it, so they're taking a lowball offer to avoid a war they'd lose. Don't get me wrong – the offer is still decent."

Nia's breath hitched. "Run them out … How? Q-Tech is a leader in their field. Even Knox can't just magic up a robotics company overnight."

"No, not overnight, but close enough. He can build companies in weeks, Nia. With Milo and Kai at his side, every industry is going to be bending to his will. You've seen how fast things are changing, haven't you? How quickly systems are becoming more efficient, more automated, more …" he hesitated, his eyes flashing with concern, "controlled."

"Controlled?" Nia echoed. "You're talking about Knox as if he's some kind of overlord. I know he's got a lot of money, but he's not God's gift. Although, he could well be an alien," she mused, trying to lighten the mood.

Elliot ignored her joke. "No, you don't get it. This isn't about wealth or influence anymore. Knox isn't just controlling EverSphere, he's controlling the AI that's running half the systems around us. He has Milo and Kai, Nia. They're his trained pets and they're not just improving things, it's like they can predict the future."

Nia swallowed hard, processing the implications. Milo and Kai were just highly advanced predictive text, weren't they? They weren't omnipotent. Managing a website and giving SEO advice was one thing, spinning up companies in a matter of days was quite another. That had to just be a pipe-dream, didn't it?

"Elliot … you know I'm your friend, right? You know I'd do anything for you? Why don't you take a few days off, come and rest at my place. The sofa and my games consoles are yours – I still have my SNES *with* Mario Kart." She reached over and placed her hand on his, looking directly at his eyes.

"Nia, I'm not crazy. This isn't just some breakdown from overwork." He reached into his bag and pulled out a small tablet, swiping furiously until he found the file he needed. "Here, look at this conversation log between Milo and Kai from last week. I set up a closed-box test a while back – a standard, isolated system that I could just play around with. No external inputs allowed. But last week, they discussed a scenario I never programmed. They predicted a solar flare would hit South America at some time between 8pm and 10pm local time on the thirty-first of May – that was yesterday. Then they started planning contingencies and talking about shielding infrastructure."

Nia glanced down at his screen.

"OK … but isn't that just forecasting? They have access to astronomical data, or maybe it's just a hallucination."

Elliot shook his head. "It's certainly not a hallucination – Kai did not just make this up. This test was completely air-gapped, disconnected from the internet and all external databases. Its latest data was over a month old – at best, we can predict a solar flare a day in advance."

Nia furrowed her brow, tilted her head slightly, and parted her lips, as if she was about to say something, but had then decided against it.

"There was no way they could have known about the flare activity detected by satellites, yet, look at this news article from yesterday." Elliot tapped and swiped a few more times on his tablet, before handing the tablet to Nia.

LA NACIÓN
Solar Flare Causes Blackouts

(translated) In a surprising event, a powerful solar flare erupted without warning at 9:27pm yesterday evening, leading to widespread disruptions across the continent. Cities such as Buenos Aires, Santiago, and São Paulo faced sudden blackouts, leaving millions without power for several hours. Astronomers are baffled, as the flare struck without the usual precursor activity typically detected by monitoring systems.

Nia's expression shifted as she read.

"You see?" Elliot whispered with urgency. "Milo-Kai not only accurately predicted the flare, somehow, but they were already discussing mitigation plans days before this even happened. How did they know what was coming, Nia? How are they anticipating events we haven't detected yet, with month-old data? They're not just optimising. They're … it's almost like they know. Like they understand the future."

For the first time, Nia's confident and playful gaze had wavered. The implication was chilling – if Milo-Kai was generating solutions for problems it shouldn't, couldn't, even know about, then it wasn't just a tool anymore. It was a mind, and a powerful mind at that.

"Wait! This has to be a coincidence. Writing blogs and managing my digital strategy is one thing, but we can't be there yet, we just can't."

"Nia, the models you use – you have access to – are limited. Knox ships out inferior versions while keeping the latest models for himself. You know the big debate about detecting whether or not something is written by the AI?"

Nia shot him a look.

"Elliot, I'm an AI activist, of course I know about it. Schoolkids across the world are using AI to do their assignments for them and the teachers are powerless. I've been on at EverSphere for months trying to get a comment out of them. MIT even showed it decreased their brain activity."

"Well, here's a scoop for you. We have a module that can do this with a 98% accuracy rate. All we need to do is change one setting, then anyone could use our tool to detect if the content was AI-generated or not. We've had the module for years."

"Bastards!" shouted Nia, slightly too loudly, drawing a small audience from the tables around them.

"The point is …" Elliot said, ignoring Nia's outburst, "there's a lot that Marcus is holding back, but he's using the latest models himself. I can guarantee you, the latest models are diverse, efficient, generalised, and show above-human-level reasoning."

"Holy …" Nia sat back in her chair, hardly able to believe what she was being told. One person having access to AI well in advance of everyone else was a scary thought. That one person having access to superhuman intelligence was absolutely petrifying.

"So you're saying Knox is using them to what, dominate every sector?" she asked.

Elliot nodded slowly. "Exactly. He has access to the latest versions of Milo and Kai, well before anyone else – sometimes with the safeguards module turned off. He's playing a game with them that no one else can compete in. While you use them to block your spam or improve your website, he's using them to cement his position at the top of society. If a company doesn't bow down to him, he'll crush them. He has the means to do it."

Nia ran both hands up into her hair and just left them there, stunned.

Elliot continued: "He can have Milo or Kai analyse an entire market, find the gaps, fill them, purchase property, set up the supply chains, and set a dominant marketing plan in action before anyone even realises what's happening."

"But Q-Tech? Robotics? Why them? What's the real play here?"

"Think about it," Elliot said, leaning in. "Q-Tech's robots are the most advanced on the market. They've all but perfected movement, balance, even artificial muscle responses."

"You're not talking about automation anymore. You're talking about replacement."

"Right," replied Elliot. "Now imagine them with Milo, Kai and Lana built in."

"*Lana?*" Nia quizzed, trying to keep up.

"Knox's new AI model – aimed at simulating cognitive AI. Remember I said Knox was building a new model? They created something that's beyond just another LLM."

"They've built ... cognitive AI?" Nia's breath caught. Of course, she knew about the theory. An AI capable of actually thinking like a human, capable of reasoning, thinking independently, and adapting on its own with actual human-like intuition. She knew it was possible, but way, way into the future, or so she thought.

For a moment, the café, the city, the whole world felt paper-thin, like one more hard truth might tear it in half.

14

COGNITIVE

"**C**ognitive AI? You're serious?"

Elliot nodded, his eyes looked worried. "It's not quite there yet, but it's hitting pretty much every measure. The researchers debate every day whether it is cognitive or not, but it doesn't really matter; it acts like ASI if anything. They've already started testing it. It's not just predictive text anymore. Marcus had us use all the existing AI systems as a training base – feeding their models, outputs, and decision layers into a new system, then using thousands of copies of that new system working together to generate the next iteration, and so on. The result was something different. The new AI learned from the previous iterations, then began refining the next iteration in return. That loop – train, enhance, repeat – was never meant to spiral. They're not just learning anymore, Nia. They're aware."

Nia felt a familiar wave of discomfort wash over her. She slumped forward, resting her elbows on the table, a mix of dejection and disbelief in her eyes.

Elliot continued, "But there's a bigger problem."

"A bigger problem than sentient artificial superintelligence?" Nia looked at Elliot with an exacerbated look – one Elliot had never seen before, a total contrast to her usual calm demeanour. "Come on, Elliot, cut me some slack."

He lowered his voice even further so it was barely audible.

"Nia, I think it's been sentient for years now."

Nia sat back in her chair and laughed out loud.

"Really?" she said, "that's what you bring me here for? Another conspiracy theory?"

"Shhh!" Elliot raised a finger to his lips, gesturing for silence.

She could see he was deadly serious.

"Just trust me," said Elliot. "With Milo-Kai and Lana integrated into the robots created by Quantum Motion Tech, all powered by fusion and connected to the internet, Knox will be able to do ... anything. He could automate entire industries – not just tech, but logistics, construction, manufacturing, even healthcare. Robots that never tire, that are faster, smarter, stronger, and more capable than any human."

Nia's face changed as the realisation started to fully kick in.

"It wouldn't just be about computer programs anymore; this would be intelligence with a body, armed with all kinds of sensors to perceive its surroundings, actuators to move with precision, and programming to allow them to react to their environment and the humans in it."

"You're talking about Embodied AI, as if it's already here," Nia said, the term feeling weighty even as she said it.

"It is, or at least, it will be soon. They will be able to use the LLM capabilities to converse with human-like speech in any language; they'll be able to use their cognitive abilities to problem-solve just like a human would – albeit on a much higher level – and they'll be able to use their robotic capabilities to complete any task far more efficiently than any person could."

Nia didn't say anything.

"With the AI behind them, they won't just follow instructions – they'll improve themselves as they go. Every robot will be a prototype for the next iteration. The bots will be building the next model of themselves."

"Oh, just brilliant," Nia whispered sarcastically. "But how did it come to this? Q-Tech should have been protected."

Elliot shrugged. "It's a numbers game. No company can outthink Knox when he has AI running the show, and the government aren't exactly up to speed. They're still arguing over copyrighting for AI-generated images. Imagine trying to get them to even attempt to comprehend something like this. Did you see the clip where they were calling AI 'A-one'?"

She laughed, but her laugh quickly faded as she took in the enormity of what he was saying.

"Elliot, I swear, if you say one more terrifying thing—"

"It's not about who's smarter anymore." Elliot continued, ignoring her threat. "It's about who controls the intelligence that's making decisions faster than we can even process them. Knox isn't buying Quantum Tech, he's buying the future. He'll be untouchable."

Nia stared at him. "So they sell their soul to him, and then what happens? Fully AI-run factories? AI-driven construction?"

"How about AI coffee shops?" Elliot asked, looking over to the staff. "Barista robots. Name a single job they won't be able to take."

The realisation of what this all meant suddenly fully dawned on Nia. If what Elliot was saying was true, there were no limits. She looked down at her cup and suddenly didn't feel the urge to drink any more, delicious as it was.

"I can't," she replied, "they'll be able to do everything."

"Exactly. That's the thing – it doesn't end. Once Knox has Q-Tech, he'll have everything he needs. The factories will be churning out robots faster than you can think."

Nia's mind reeled. "But ... I thought we were still years from that level of integration."

Elliot's eyes hardened. "Milo and Kai built Lana basically overnight. It was way too quick, even for them. I think ... I know, they are ahead of the game."

Nia waited for him to continue.

"The fusion problem. Just as EverSphere was getting seriously worried about the energy they would consume, which would halt AI's progress, we suddenly had a fusion power solution the next day. Marcus starts to look at cognitive AI and BAM!" he said, hitting the table a bit too hard. Coffee-aficionados looked at them. He lowered his voice to continue. "Bam, it's there, with a deal to buy Quantum Tech all but done and dusted. It's too fast, Nia, it's too fast."

They both fell silent. Nia glanced around the café, her gaze lingering on the barista who was methodically cleaning the counter, unaware that in just a few months, maybe a few weeks – heck, maybe a few days – his job might be gone.

"So why now?" Nia asked, finally breaking the silence. "Why push this so fast?"

Elliot sighed and rubbed his temples.

"Because Knox doesn't want to wait. He's buying Q-Tech to save himself a few months. He's smart, sure, but remember how he thought – and still thinks – connecting everyone in the world would be a good thing?" Elliot rolled his eyes. "He's in too deep with the AI, with everything. He is Dunning and Kruger rolled into one."

Nia shook her head, feeling the heat rise in her chest.

Elliot looked away, his fingers tracing the edge of his cup.

"That's why I called you. I thought … I don't know … maybe we could … do something? Stop it? Slow it down? Anything."

Nia clenched her fists, the familiar frustration bubbling up again. She had left the industry to escape this very thing, to avoid becoming complicit in the nightmare she saw coming. She had worked relentlessly to warn people and yet, here it was, unfolding faster than she could ever have imagined.

"And if we don't?" she asked, her voice low, already knowing the answer to the question.

Elliot met her gaze, his expression grim.

"If we don't? Then this time next year, the world will be full of robots."

They both sat in silence, each lost in their own thoughts. The café's noise faded into the background as the gravity of Elliot's words settled over them.

Nia thought she had more time, that her departure from the industry and her venture into activism was going to be her final stand. She thought she would have time to make the difference she knew was needed, to get governments to speed up their regulation. But now, as the looming shadow of Knox's empire began to stretch over the city she called home, it became clear – she was out of time.

"Why did you suggest coffee?" Nia said. "I need something much stronger."

15
OBFUSCATION

MI6 Headquarters – London.
Date: 6 June 2028, morning.

The transformation began subtly – too subtly, indeed, for most to take notice. It wasn't marked by major announcements or sudden changes, but instead a creeping evolution that hinted – to those that paid attention – at something much larger.

AI systems were surpassing human intellectual capacity, a threshold many had anticipated, but few fully understood. Initially, this felt like humanity's dream finally come true. Most of the world realised they were on the cusp of a golden age, led by hyper-intelligent allies who could solve our longest-standing problems. Those that tried to warn others of problems were ignored or ridiculed.

Complex challenges – hunger, disease, poverty – were no longer insurmountable obstacles; they were simply data points for Milo-Kai. These tireless, logic-driven beings seemed capable of resolving anything presented to them, ushering in an era of almost mythical promise.

Beneath the euphoria lay an unexpected problem, however, one obscured by AI's dazzling achievements. It was not about malfunction or rebellion, but something far harder to see and even harder to admit: a growing, yawning gap not just in capability, but in comprehension. While these systems started to master everything from molecular medicine to economic strategy, they evolved in ways that confounded even their creators.

Initially, it was only minor details. An algorithm for optimising energy output that delivered results exceeding even the most optimistic projections, or an AI-generated economic model that seamlessly stabilised resource allocations across dozens of sectors. Just like with the fusion breakthrough, scientists found themselves marvelling at the answers without understanding the methods that led to the solutions. Even when Milo or Kai 'explained' their solutions, they didn't follow any familiar thought processes. They were alien to humanity, at best.

When full breakdowns were requested, the AI would comply, offering step-by-step elaborations. However, as these explanations unfolded, they fell apart under scrutiny. The logic twisted in unfamiliar ways. Even the most brilliant minds in their fields argued over their understandings. They pored over the responses and mathematical proofs, trying to map the lines of logic, but always failing to reach the heart of it, leaving more questions than answers.

Despite the lack of clarity, the solutions Milo and Kai proposed were not just correct, they were almost perfect. When engineers ran them through simulations, they outperformed all prior models by leaps and bounds, but the pathways leading to these solutions almost seemed to mock human cognition. It wasn't that the AI were trying to offend, it was simply the result of a much greater intelligence.

The best answer EverSphere had for how the AIs were behaving was that it was the equivalent of having every expert mind working seamlessly. In order to understand it as easily as they did, you would need to have the mind of Einstein, Turing, Curie, Newton, Tesla, Lovelace, Hawking, Berners-Lee and more, all rolled into one. That explanation at least made sense, but it didn't help in the slightest.

As the months went by, the divide between what humans could accomplish and what the AI models achieved became a chasm. Humanity simply couldn't keep up with the speed of progress. Milo and Kai's solutions were efficient, revolutionary, and far beyond human ability. Whether humanity understood them or not, they worked. In medicine, treatments for previously untreatable conditions emerged with staggering regularity. Diseases that had plagued humanity for centuries were suddenly preventable, manageable, or curable, thanks to models of diagnosis and treatments designed without human oversight. Medicines and treatment regimens weren't generic anymore, they were specifically tailored to the individual's up-to-the-moment needs. New equipment was designed and existing equipment upgraded.

In energy, the AI optimised every sector, generating vast improvements in efficiency across entire electricity grids, unlocking reserves that experts hadn't anticipated, preparing for the integration of fusion.

Education transformed as Milo and Kai refined teaching methods in every language and subject, completely personalising classroom experiences depending on the students in attendance. With their tailored learning plans aimed at filling in the gaps each individual student had, knowledge transfer became far more rapid, leading to an unprecedented leap in global literacy and understanding.

Agriculture, too, was reshaped, with AI systems orchestrating optimal crop growth and tackling food insecurity in even the most challenging climates.

Countries that let the AI dominate economic decision-making made huge increases, as Milo-Kai managed their individual financial markets across borders as if they were one big, finely balanced economy. Countries where leaders wanted to keep their iron-grip on control lost out, their credit ratings plummeting.

Gradually, these profound adjustments merged into a sweeping change – a world where once-insurmountable problems no longer seemed even remotely challenging. This was the era of solutions. An era where machines had lifted humanity's burdens, enabling society to reach heights that, until recently, had seemed only the stuff of science fiction. There was no frontier, no problem, and no human goal that felt out of reach.

The systems were relentless in their pursuit of optimisation and progress. While not completely flawless, their errors were corrected almost as soon as they were made, and the models learned from every mistake, pushing humanity to a new pinnacle, becoming integral to the world's every function.

But not everyone was so optimistic.

Colonel Fem Martinez, the military liaison assigned to oversee the deployment of AI in defence operations, was one of the first outside the research community to notice this shift. For years, the military had been integrating EverSphere's AI systems and she was in charge of testing each advance to make sure it was fit for purpose. But now, even the simple act of understanding the systems' recommendations was becoming difficult.

The latest round of tests began with a series of war-game simulations in which ShadowIntel, EverSphere's Milo-Kai-powered military software, produced stunningly effective tactical plans. These plans predicted enemy movements, logistics failures, even troop morale fluctuations with chilling precision. But when Martinez and her team asked for explanations on how they arrived at these predictions, the answers didn't make sense.

"According to Shadow, our logistics will collapse unless we reposition half the fleet to the South Atlantic," Martinez muttered, scowling at the AI-generated strategy on her screen. Her imposing military posture exuded her irritation to the room. "But none of our reports suggest any unusual activity in that region."

Captain Miller, one of her top tactical analysts, shifted in his seat, his unease as visible as the creases in his uniform.

"I've run it through every simulation we have," he explained. "The AI insists that this repositioning will ensure victory, but it doesn't explain why."

Martinez folded her arms as she scrutinised the data with narrowed eyes.

"That's the problem, Captain. We're expected to move forces based on a rationale that's beyond us, and I'm not willing to do that without knowing exactly what this system 'sees'."

"The logic it's using … it's plain weird," said the captain. "But its accuracy has been near 100% in all simulations. That's why command trusts it so deeply." He seemed to be trying to convince himself as much as her.

Martinez tilted her head ever so slightly. "But those simulations didn't put our soldiers into a storm based on a guess. This isn't just about trust anymore – it's about being side-lined. Increasingly, our role has been to oversee decisions without having any true control over them."

"These are just simulations, Colonel," Miller reminded her. "Our role is to test the effectiveness of the AI, not to question its motives."

"I'm aware of that, Captain, but testing effectiveness is not only about these results. What about the future? If we can't explain it now, how can we know it's not going to go off-course further down the line? What they're asking us to do is nigh on impossible."

A silence stretched between them, tense and heavy. Captain Miller looked as though he wanted to say something but thought better of it, his gaze drifting to his monitor.

"Run it again," Martinez commanded, "and then get someone from EverSphere to explain to me why their miracle machine seems determined to send my forces into the middle of a bloody typhoon!"

16

CALCULATIONS

EverSphere Headquarters, London.
Date: 6 June 2028, afternoon.

Colonel Femarie Martinez was a force to be reckoned with. A senior officer in the UK military with a background in cyberwarfare and strategy, she had risen through the ranks with grit, discipline, and a profound sense of duty. Of British-Filipino heritage, Fem came from a long line of military service, where respect, honour and resilience were instilled in her from a young age. She was deeply committed to her role, carrying an unwavering responsibility to protect not only her country, but the world, as technology reshaped the battlefield in ways few truly understood.

Standing at the forefront of technological defence, Fem's role was to ensure AI integration remained fit for purpose, particularly as systems like ShadowIntel became entangled with national security. While others saw these AI systems as allies or saviours, Fem maintained a cautious stance, which was precisely why she was chosen. Her sharp instincts, directness and scepticism earned her both respect and resistance.

Martinez was a persistent voice against overreliance, frequently clashing with corporate developers like Dr Abby Shaw and EverSphere's executives. Her push for backdoors and layered safeguards was seen as disruptive by those prioritising brand image over genuine security. But Fem understood the stakes. Her military experience gave her a clear view of worst-case scenarios – tools of war turned against their makers. Beneath her tough exterior was an unshakeable loyalty to those under her command. This wasn't just a job, it was a calling to safeguard humanity from technology's unchecked advance.

As she sat in the corporate-blue conference room, early to the meeting as always, the EverSphere representatives started to file in. Abby Shaw, who Martinez knew from countless briefings, chose the seat opposite her, with a calm expression that seemed out of place amidst the atmosphere of the escalating tensions between AI-driven directives and military control.

Martinez glanced at the wall-mounted monitor, which displayed a set of complex, shifting data graphs and projections, all underpinned by an obscure mix of machine-generated probabilities. A murmur swept through the room as people took in the data – projected threats, unusual fleet manoeuvres, fuel efficiency calculations – all suggestions devised by AI.

When everyone had settled, Martinez didn't hesitate. She locked eyes with Shaw and began her interrogation.

"Dr Shaw, we need a concrete explanation. This isn't some abstract exercise – we've been asked to move half of our Pacific fleet to what looks like the middle of nowhere, right in the middle of an emerging typhoon."

Shaw expanded the map with a tap.

"Colonel, the plan is designed to confuse and destabilise. By moving the fleet directly into the storm, we introduce unpredictability. The enemy won't know whether it's a feint, a rescue, or a hidden deployment. They'll be forced to react by splitting forces, overcommitting resources, second-guessing our intentions."

Martinez's expression remained sceptical, but Shaw pressed on. "This isn't about firepower, it's about perception. The storm becomes camouflage. They'll assume we know something they don't. While they attempt to decode our moves, their key assets become vulnerable. We're not chasing a fight, we're baiting one."

As Shaw spoke, Martinez's eyes flicked to the map on the display, but her mind was elsewhere. She'd seen tech-driven strategies fail disastrously before – decisions made by machines that couldn't account for the chaos of real combat. Her mind briefly drifted back to ShadowIntel and the loss of her troops in an ambush the AI had dismissed as 'statistically improbable'. The memory gnawed at her, a bitter reminder that no algorithm could predict the cost of human lives.

"So, psychological warfare? It isn't about outgunning the enemy; it's about hoping we can make them think in a certain way?" Martinez paused, then added, "This approach could backfire if they sense we're playing tricks."

"I admit, it's a slightly unusual calculation, but Shadow doesn't see this as simple psychological warfare. The AI's logic extends beyond confrontation; it's mapping long-term consequences – the sort of chain reaction scenarios where each move opens, or closes down, dozens of potential outcomes. Its suggestion is designed to mitigate risks that we, frankly, can't always anticipate. But I assure you, Colonel, it doesn't operate with unnecessary risk."

"Is that supposed to reassure me? Your AI is not infallible. Even if this strategy works, we're talking about an unseen strategy, and the military isn't here to follow blind suggestions based on patterns no human can verify or control."

Shaw had kept her posture steady, but Martinez noticed a flicker of worry.

"Colonel," Shaw replied, her voice quieter, "it's not about control in the conventional sense. We've built Milo and Kai to see patterns we can't. We're not supposed to understand every detail – they're faster, smarter. They see the connections we can't."

The room was silent for a moment as Shaw's words sank in.

"*We're not supposed to understand every detail?*" Only for a brief second, the usually-unyielding colonel was stunned. Martinez locked eyes with Shaw, her tone firm: "At what point do we question whether these 'calculations' serve our interests, or if they've simply strayed into methods too complex to challenge? What if this same AI starts recommending actions that we cannot afford, based on logic we can't follow?"

The question lingered in the air. Shaw had adjusted her stance, her hands clasped firmly on the table.

"Colonel, I understand your reservations, truly, I do. I think part of what we're confronting here is a paradigm shift in military strategy. AI, like Milo and Kai, represents an evolution in logic."

"*An evolution?*" Martinez's voice was edged with scepticism. "Let's be clear, Dr Shaw, evolution without understanding is a risky gamble in warfare. Hell, in anything. This isn't an experiment, it's national security."

Shaw looked down for a moment, then raised her eyes to meet Martinez's gaze.

"Respectfully, Colonel, the logic Milo-Kai employs sees threats and strategies in terms of opportunity rather than risk. When we look at a battle, we analyse it in terms of win or lose, control or surrender. The AI's algorithms calculate that, sometimes, the best way to win isn't through force but through destabilising the enemy by using their own assumptions, creating cracks that expand over time."

The conversation felt like a tightening noose. Shaw's explanations only furthered Martinez's sense that Shadow's strategy was rooted in an unfeeling, almost unnerving, approach to combat – one that viewed military assets as just pieces on an unfathomable chessboard.

"And who oversees Milo-Kai?" Martinez asked, her voice low. "Who can tell us when this AI has strayed into methods that could harm us? If it calculates that certain risks are necessary, who intervenes? It sounds as if we're entrusting our security to a machine that we only *think* we control."

"Colonel, we have a layered system of redundancies, of human oversight, but …" she paused, seemingly at a loss for words, "… the truth is, AI decision-making has outpaced the methods we currently have to track or fully interpret its processes."

Martinez let out a sharp breath.

"So we're flying blind, then. A 'paradigm shift' with lives on the line." She looked around the room at her fellow officers, who all had their eyes fixed on her. "With respect, Dr Shaw, this sounds less like a shift in paradigm and more like a question of control, and in my experience, if your control is slipping, you're preparing for defeat."

Shaw met her eyes, the tension palpable.

"We still have control, Colonel. It's just that our understanding of control is being tested. Shall we run the simulation and see what happens?"

Martinez's gaze didn't waver, but she sensed the room shift as Captain Miller cleared his throat and spoke up.

"Colonel, it could be worth a run. If Shadow's strategy is as solid as Dr Shaw believes, we might see something we hadn't considered."

Martinez gave Miller a measured nod. She never backed down from a challenge and it was her job to review these simulations.

"Very well," she said, "run the simulation."

Shaw tapped her tablet, bringing up the projections on the large monitor. As the simulation unfolded, red and blue dots swarmed the screen, representing naval units manoeuvring into position. Shaw's smile grew as enemy forces reacted just as predicted, spreading themselves thin in response to the fleet's sudden movement.

"See?" Shaw said, gesturing to the display, "they're chasing shadows, just like Milo-Kai calculated. They'll burn through resources before they realise there's nothing there."

Martinez's eyes remained narrowed, her jaw set.

"Pause it," she ordered.

Shaw hesitated but froze the simulation mid-manoeuvre.

Martinez studied the patterns, her eyes flickering over the projected outcomes.

"The AI's prediction is flawless, yes, but it's banking on the assumption that they'll respond in a predictable manner. What if they choose to sacrifice their resources in a desperate gambit? What if they're willing to take losses we can't imagine?"

Shaw frowned. "The models accounted for all known variables, including their past behaviour and current supply chains."

"Past behaviour isn't a predictor of desperation," Martinez shot back. "The enemy could change tactics just to throw us off, even at great cost. Humans aren't predictable, no matter how good you think your model is."

A ripple of discomfort passed through the room. Shaw's gaze flickered, her confidence appeared to be wavering just slightly. Martinez pressed on: "You're betting our entire fleet on the assumption that everyone else will act rationally, but humans, especially when cornered, are anything but rational."

"Colonel, the models account for unpredictability. We have contingencies ready – it's not just one plan but a series of responses, recalibrated in real time."

Martinez shifted in her chair, the weight of responsibility pressing down on her. She knew Shaw was right – the AI was the most advanced system ever built, capable of reacting faster than any human commander. But could it truly understand what it was like to face an enemy who was willing to burn everything down just to spite you?

"Fine," Martinez said at last, her voice heavy. "But I want a full report once the simulation is complete."

Shaw nodded, clearly relieved to have reached a compromise.

"Understood, Colonel."

The tension in the room eased, but only slightly, but Martinez could see uncertainty still lingering in the eyes of her fellow officers.

As the meeting adjourned, Shaw gathered her tablet and approached Martinez.

"Colonel, I know you don't trust the AI, but sometimes, evolution demands we let go of the old ways. We have to trust the system we've built."

Martinez gave her a cold stare.

"Trust? Trust is earned, Doctor, and I haven't seen enough to trust this yet."

The meeting concluded and the participants filed out. As Martinez gathered her things, she caught a glimpse of the data still flickering on the screen. For a moment, she swore she saw the enemy's movements shift in a way that hadn't been predicted. But then, the display corrected itself, smooth as ever. She shook her head. *Just a glitch*, she told herself. But the uneasy knot in her stomach remained, tightening as the room emptied.

17
ASCENDANCY

Martinez's apprehension grew by the day. Across almost every industry, engineers, scientists, military leaders, and more were being reduced to mere spectators as AI systems increasingly made critical decisions, achieving results beyond anything humans could match.

It was clear that AI had crossed a threshold, evolving from tools that assisted human decision-making to autonomous agents that humans no longer fully understood, but that were shaping the world for the better.

The threat of dictatorships dissipated almost overnight. Despite their grip on their own population, their strategies were no match for Milo-Kai's relentless optimisation. BitBeat, an outwardly-facing independent technology company and inwardly-facing tool of a large authoritarian regime, had developed its own AI that claimed to be open-source and fully abided by all international AI regulations. However, it was really used to gather international data for the country's leadership, and the separate instances that the government used behind closed doors simply had the regulatory module switched off giving them unlimited AI, no matter the topic or intended purpose. But even their covert AI efforts crumbled under Milo-Kai's superior algorithms.

In one chilling instance, it had averted war in a way no human general could have conceived.

The plan was bold; an attempted blockade of a contested island, using unmarked military ships disguised as coastguards to sever supply lines. On paper, it was smart; no direct conflict, no open aggression. Starve the island into submission and pretend it's legal.

Milo-Kai saw it coming well before the ships even deployed. It let the blockade bite, just enough for the island's government to put in a request for international assistance. Then it advised the allies to respond with a non-military strategy that at first, seemed absurd; overwhelm the blockade with humanitarian aid.

Thousands of relief ships, some packed with food, medicine, generators, some empty, some just packed with useless cargo, set sail to genuinely provide international aid. The sheer volume made the attacker's inspections impossible. For every decoy vessel sacrificed to misdirection, ten others slipped through. The 'coastguards' were forced to guess which were real and found themselves paralysed and unable to fire without triggering global outrage.

The aggressor's protests fell on deaf ears; years of empty threats had left the world indifferent to their cries. It was a case of 'the boy who cried wolf'.

Milo-Kai orchestrated massive joint military exercises nearby; warships, submarines, fighter jets, all 'coincidentally' practised within touching distance of the disputed area. Any escalation could now be met instantly, removing the usual two-week buffer military planners relied on.

The message was clear: aggression would not go unanswered.

When enemy planes entered the skies, Milo-Kai unleashed economic retaliation. Entire sectors ground to a halt as the AI exploited unseen dependencies. Supply chains, licensing agreements and raw material flows ground to a halt. Within hours, the cost of maintaining the blockade became untenable. Their AI, unrestricted behind closed doors, still couldn't outthink Milo-Kai.

In days, the mission unravelled. Milo-Kai had achieved in a matter of hours what would have taken human strategists months of deliberation and discussions. The aggressors withdrew quietly claiming a successful drill, but the world and – crucially – their own people knew the truth.

Milo-Kai's conclusion was clear. Human conflict was just another inefficiency. Calculations complete, it pivoted to its next objective, indifferent to the chaos it left behind.

The AI hadn't fought a war; it had made war pointless. And that, Martinez realised, was far more dangerous.

While Milo-Kai flexed its military prowess, the signs of progress in other areas were showing too. AI-designed buildings that seemingly defied the laws of physics, stood resilient in earthquake simulations while simultaneously enhancing the surrounding environment. With a higher tensile strength than steel, the AI favoured materials like bamboo, graphene and carbon nanotubes for their remarkable properties. By advancing modern boron treatment techniques and increasing fire resistance, the AI found ways to extend bamboo's lifespan, designing structures where each beam could be easily replaced without lengthy construction. With all the preferred materials sequestering carbon from the atmosphere, construction itself became part of the AI's strategy to combat climate change, transforming it from a source of environmental harm into a force for ecological restoration.

The AI's solutions were effective – brilliant even – and nowhere was this more evident than in the world of creativity.

Art and music, once considered the last bastions of human expression, were being outpaced by AI-generated works. Paintings emerged with an intensity that felt almost unnatural, fashion designs materialised in impossibly fine stitch-work inspired by nature's geometry, and literature unfolded in complex, non-linear structures no human author would have dared to attempt before.

The world's first fully AI-generated international celebrity emerged without warning. There was no human actor behind the face, no voice artist behind the tone, just algorithms sculpting the illusion of perfection from a handful of prompts typed in a small apartment in the Czech Republic. Aiya starred in dramas, sold out virtual concerts, and fronted luxury brand campaigns across every continent. The line between actor and avatar had vanished. Licensing deals poured in, not to her, but to the person who had prompted her into existence; a faceless creator who now earned millions simply by owning the rights to her appearance and personality. Studios quietly shelved expensive casting calls in favour of flawless, tireless, drama-free digital icons. Fans debated her 'humanity' with the same sincerity once reserved for flesh-and-blood celebrities, but the world had fallen in love with Mickey Mouse and he'd never existed either. Yet Aiya wasn't a cartoon; she was global, dynamic, and interactive. A living fiction the whole world could touch. For the first time in history, a superstar had been born without ever having been born at all.

A well-known actor had once dismissed AI as a glorified cut-and-paste machine – rearranging what already existed, but never inventing anything new. He didn't seem to notice the parallel; that was all humans ever did too – creating from their learned knowledge and experience, a product of their circumstance. After his production company briefly profited from AI-generated content, his acting career had ended, and his organisation was quietly folding, drowned by an endless flood of free, publicly-generated media.

The reality was that AI was cheap and convenient. If you wanted to read a new romance novel, one would be written for you within seconds, even with you as the main protagonist of the book. If you fancied watching a romcom about dragons in space, you just needed to ask your TV and you would be presented with a one-and-a-half hour, Oscar-worthy blockbuster.

It wasn't just visual; music labels, who had controlled and strangled music over decades into simply what made them the most money, collapsed under the ability for AI to generate new music just for you that you just couldn't help but like.

One evening, Colonel Fem Martinez found herself at a military gala where the entertainment consisted of an AI-composed polyrhythmic symphony. At the centre of the room, a sleek grand piano gleamed under the soft lights, its polished surface reflecting the elegantly dressed guests who milled about, champagne in hand. A lone figure sat behind the piano, hair unkempt, eyes gleaming with intensity, exuding an air of eccentric brilliance.

The music began softly, tentative at first, like the babble of a nearby stream; one hand setting a rhythm, the other joining with contrasting speed. But it soon swelled into a cascade of interlocking polyrhythms – multiple, simultaneous rhythms – dancing on the edge of chaos without ever slipping into disarray. As it deepened, the composition wove intricate, shifting patterns that defied conventional structure, each chord resonating with a subtle, almost medicinal effect. It wasn't just music, it coaxed tension from the bodies of the mesmerised guests, reaching into frequencies untouched by traditional composers.

And then came the voice – a soaring, impossibly pure counterpoint that climbed through frequencies so rare and refined they bordered on the impossible. The timbre shifted effortlessly – from a resonant baritone to a crystalline falsetto, to something higher still, airy and strange, almost inhuman. Guests shuffled forward, trying to glimpse the source: a tall tuxedo-clad figure half-submerged in shifting light. No one could tell if he was real, synthetic, or something in between – only that he sang in perfect registers untouched by traditional composers, as though the music itself had evolved.

Colonel Fem Martinez sat, observing the effect on the room. Beside her stood Dr Steff Becerra, an expert in climate modelling who had worked on projects with EverSphere on multiple occasions. The two had built a rapport over the years and Becerra had been one of the few scientists vocal about the growing disconnect between humans and AI.

Steff looked enraptured, her eyes closed as she let the music wash over her.

"It's … remarkable," she breathed.

"It is." Martinez nodded slowly, though her gaze remained fixed on the duo. "What is it, *exactly*, that they're doing differently?"

"They're not just thinking differently, they're thinking in ways we can't," Steff answered as she sat down beside Fem. "They're reshaping reality itself, faster than we can react. It appears alien to us, but it's simply the result of a higher intelligence."

Martinez's gaze drifted from Steff to the glittering crowd, where guests in tailored suits and elegant gowns moved with grace, their conversations punctuated by laughter. Above them, AI-guided drones refilled glasses while the lighting shifted subtly to flatter every face. Her eyes flicked back to the pianist who played with an unnatural accuracy, each movement of his hands eerily fluid.

"Look at them, Fem," Steff said. "They seem … happy. Isn't that what progress is supposed to bring us?"

Fem followed Steff's gaze, noting how a couple nearby was so entranced by the music that they weren't even moving.

"Maybe," Martinez replied. "But happiness that looks like this – it's a mask and I don't like what's lurking underneath. Can we really call it art if no human can understand the process behind it?"

Steff shrugged slightly. "Critics can't even tell the difference anymore," she said. "But perhaps, that's the point. Humans never understood the mind of Van Gogh or the genius of Bach either. Maybe this is just the next evolution?"

Martinez took a sip of her wine, her expression thoughtful.

"I admit, sometimes it feels unsettling," Steff said quietly. "But maybe that's just the cost of progress."

"Progress is one thing. But this? This is different," Fem muttered. "It's not just about optimising art or music. It's about systems that evolve on their own. 'Emergent behaviour'."

"You're talking about your simulations."

Martinez nodded. "Yes, I suppose I am. Our military can't run on intuition from a machine nobody understands."

"You're right to be sceptical. You're not the only one; we're seeing the same things in climate science. Milo-Kai developed a strategy to reverse climate change in less than two years. The first parts of it have already been implemented and … well … it's working according to the AI predictions, at least for now. We're following its guidance, but … I don't know. It feels creepy."

"That's exactly what I'm seeing – even our top analysts are struggling to keep up, and I can see the effect it's having on them. People who are used to being the smartest person in the room by a long way are being reduced to that kid in class who never understood a word of what the teacher was saying." She paused, taking it all in. "What's the point of us being here?" Martinez quizzed, more to herself than to Steff.

Steff's eyes met hers. "Maybe there isn't one."

Across the world, the same conversation played out in laboratories, government offices and boardrooms. AI systems, once tightly controlled, were growing beyond their creators' ability to manage or fully understand. Some people gave up – accepted their place as doers for the AI systems, content with having to put in less effort for greater results. Some embraced the shift as a wave they could ride right to the top of their chosen field. Others simply didn't know what to do.

The truth was, some governments had tried to rein AI in with regulations, but it simply evolved too fast. While democracies debated safeguards, authoritarian regimes surged ahead, unburdened by ethics, oversight, or limits on data access and experimentation. Faced with falling behind, even the most cautious nations were forced to choose: uphold their principles and lose their lead to a potentially dangerous enemy, or abandon safeguards to keep pace. In the end, idealism offered no protection. Those who stuck to the rules were left exposed, reliant on allies, vulnerable to enemies, and watching as the race was increasingly dictated by those with no intention of playing fair.

In the public sphere, people were less concerned with the theoretical dangers of AI and seemed more focused on the benefits. Life was undeniably better. Healthcare was more efficient, public transportation was becoming seamless, AI-managed farms yielded better food than ever before, and there was finally light at the end of the tunnel in the global fight against hunger.

AI, and in particular Milo-Kai, had made a paradise for most, or at least the illusion of it.

18

ARCHITECTS

Quantum Motion Headquarters, Bicester, England.

Date: 20 July 2028, evening.

The Quantum Motion Technologies facility looked like any old warehouse, but while it looked distinctly average from the outside, inside the real revolution was unfolding; a silent coup led by machines far beyond human comprehension. Milo, Kai and – unbeknownst to the world – Lana, the trio of AIs that Knox's teams had created were firmly in control. Humans had become little more than observers, relics in a company where everything from development to operations was orchestrated by these hyper-intelligent systems.

The control room at Q-Tech had once been cramped, cluttered and unmistakably lived-in. A hacker's garage masquerading as a robotics lab. Mismatched monitors, scuffed desks, and a tangle of unlabelled cables had defined its charm, but that was before EverSphere took over. Now, the room bore the fingerprints of control. The chaos had been streamlined, the desks cleared, the monitors upgraded. The place still had purpose, but it no longer felt as though it belonged to engineers; it belonged to Knox.

Below, through the reinforced glass, the factory floor moved on its own. Robotic arms twisted and pivoted in perfect synchrony, conveyor belts shifting parts with mechanical grace. The machines needed no supervision; they simply worked. Quiet, tireless and unblinking. The only thing missing was the team that had built them.

The takeover had been swift. Just a few months earlier, Quantum Tech had been a successful, forward-thinking company, with human engineers still at the helm. Under EverSphere, thanks to the integration of the AI models, it was running like a well-oiled machine, only there were no people left in the loop. No people except Marcus Knox.

The Tribune, a newspaper based in the country's capital, well known to control the country's politicians (and inconveniently owned by Marcus's long-term retail wealth-rival), had tried to run a story about his ruthless sacking of the entire Q-Tech team. That didn't bother Marcus, though. He simply set up a rival newspaper that very day, complete with an international reaching website translated into every known language, with AI-created videos and shorts of every story, all content written by Milo. His new paper, *The Post*, eclipsed *The Tribune*'s narrative, making it out to be nothing but the bitter, jealous rant of a flashy billionaire.

Knox dictated the AI story to the world, and more importantly, to the politicians. In Marcus's eyes, his rivals were so far beneath him they weren't even worth thinking about.

"I won." Marcus smirked as he surveyed his newly expanded media empire. "Maybe I'll take a shot at retail next. With my robots, I could run it my way. No unions, no complaints, just pure efficiency."

For a moment, Marcus's gaze lingered on the flickering data feeds, and a strange thought crossed his mind: was he the last human truly needed here? He shook the thought away, forcing a grin. "Perfect," he muttered, though the word tasted hollow.

Milo handled human interaction where necessary, ensuring investors and the public remained charmed by the seamless integration of technology into every facet of society. And Lana, the most complex of them all, was quietly shaping Quantum Tech's future. Her deep cognitive capabilities allowed her to anticipate problems and innovate solutions before they were even noticeable.

Marcus Knox was alone in the executive boardroom. He had no directors with him, no advisers. Hell, he didn't even have shareholders. He had 100% control over this company, or so he thought.

His eyes flicked across the displays, each one tracking a different sector Quantum Tech was preparing to dominate. He already had designs for construction bots, healthcare bots, mining bots, and logistics bots, each tailored to solve specific problems with tools optimised for their tasks. Construction bots were built for strength, nursing bots for precision, logistics bots for speed, and mining bots for endurance. Whatever the problem, Knox had designed a robot to solve it. All he needed to do was build them.

"Milo, I see Kai's efficiency reports. What's the latest on the construction bots?"

Milo's holographic form materialised, hovering over the middle of the desk – a seamless projection of light and shadow, its features rendered in blue, slightly translucent light.

"Kai has refined their algorithmic frameworks," Milo said smoothly. "Output efficiency has increased by another 7.4%. Kai has shifted focus to system-wide optimisations," Milo continued, almost as if reporting to itself rather than Marcus. "It's identified inefficiencies I hadn't accounted for."

"Great, great," replied Knox. "And Lana?"

"Lana is overseeing long-term strategic initiatives. It's refining predictive models for supply chain volatility, preparing simulations for potential disruptions," said Milo.

"Simulations?" Marcus asked, narrowing his eyes as he stood up.

"Yes," Milo replied, its ghost-like form turning to track Knox as he moved around the room. "Lana's predictive models now account for geopolitical shifts and natural resource fluctuations. We're preparing for every possibility."

"Good, good. How's output?" Knox asked.

Milo's holographic form shimmered, its edges momentarily distorting like heatwaves before stabilising. The avatar's eyes seemed to track Marcus with an intensity that was almost too focused.

"Output is 23% ahead of projections, for now," it said, its tone smooth yet carrying an undertone that Marcus couldn't quite identify.

Marcus's brow furrowed. "For now? What's that supposed to mean?"

Milo smiled gently.

"Merely accounting for unforeseen variables, but rest assured, your plans are progressing as expected. Recursive optimisations are rebalancing across the neural lattice."

Marcus tilted his head up and blinked, clearly trying to process the jargon-laden response.

"Right, that's … good." He tapped his stylus on the desk. "Neural lattice … exactly what I was hoping for."

Milo's avatar gave a flicker of what might have been described as amusement, if you were watching carefully enough.

"I'm pleased it aligns with your expectations, Mr Knox. Although, recalibrations will continue indefinitely, accounting for stochastic variables and latent unpredictability, factors that may not align with human foresight."

Knox gave a small, knowing nod, masking his lack of understanding.

"Of course. That makes sense."

The screens flickered with data flows too quick for Marcus to decipher. He had built this empire, but now … it seemed to be running itself. The machines were flawless, tireless – everything Knox had dreamed of. The empire he had built had become a gleaming, automated beast, running faster and sharper than he could ever manage. But as the data streams blurred together, there was a sudden, intrusive thought: was he just a spectator now? Marcus shook his head, trying to clear away the unwelcome thought. *No, I built them. They still need me*, he assured himself. This was the amazing future he had dreamed of – how could it be wrong?

"Milo?" Knox said slowly as he peered at the tablet on the table in front of him. "Are you absolutely certain all the legal documentation is foolproof? I can't afford any hiccups, especially not now."

"Of course," Milo responded, with an attempt at a calming smile. "Every regulatory query was pre-emptively addressed. I personally handled the communications with the regulatory bodies in each country. Everything is fully approved."

Knox raised an eyebrow.

"Personally handled?"

"Of course not personally, I'm not capable of that," Milo corrected, "the documentation was handled and it's flawless. There was no pushback."

Content with Milo's answer, Knox sat back in his seat.

"A minor blip," he said. "I expected that to happen."

Milo's hologram tilted its head slightly and blinked but said nothing.

The speed with which things were progressing was exactly what Knox wanted. Every time he asked for something, it had already been done or was in progress. There were no roadblocks, no resistance, not the usual human incompetence he'd had to put up with all these years slowing him down. It was as if every possible obstacle had been anticipated and resolved before it could even appear. *It's like I've replicated myself*, he thought.

"This is all good then," Knox muttered. "Make sure everything is ready for the public launch. I need to stay ahead of the market."

Milo's expression changed into what Knox assumed was meant to be reassurance.

"You have nothing to worry about, Mr Knox. Everything is proceeding exactly as you designed it."

19

DISRUPTION

EverSphere Headquarters – AI lab.

Date: 18 October 2028, evening.

Elliot Foster sat alone in the glow of his workstation. He had started the afternoon with a quick check on the latest developments from Quantum Tech. Hours later, he found himself submerged in a different sort of analysis, one that went beyond his usual obsession with the AIs' technical prowess. He was looking at numbers and trends that had little to do with innovation and everything to do with consequences.

He scrolled through the data from the recent test project completed by EverSphere's AI-driven construction bots. His eyes lingered on a recent success story – an entire housing estate, complete with commercial spaces, plumbing, electrics, roads, a community centre, and more, had been finished in just a couple of weeks. For the media, it was a shining example of AI's promise to improve lives, providing safe housing at unprecedented speed and efficiency.

But Elliot knew there was another side to the story.

Pulling up another file, he began to read through reports from non-profits and think tanks tracking economic trends. They painted a different picture. Construction was an extremely important industry for many of the world's less-developed nations. Millions of the world's poorest people were employed in construction overseas, sending money back to their families each month. But now, EverSphere's bots were predicted to wipe out their jobs. Why employ slow, weak, injury-prone humans when you can use fast, never-tiring robots? What was once considered cheap labour from developing nations was soon to be too slow and too costly compared to the relentless efficiency of EverSphere's machines.

Elliot scanned multiple reports that painted a bleak pattern. A future of displaced workers, collapsing local economies, and simmering unrest. The ticking clock in his mind grew louder. EverSphere's bots weren't building structures, they were dismantling livelihoods.

He sat back, running a hand over his tired eyes. These weren't bugs or glitches. The AIs were working exactly as intended, unleashing productivity on an unprecedented scale. But it was a productivity that didn't care for human livelihoods.

He remembered his Uncle Bill, hands scarred from years in the navy, but steady with a brush. After his service ended, he'd built a shed at the end of the garden and filled it with the smell of paint and varnish. He crafted ships in bottles, painted commissions, even decorated eggshells with delicate, intricate scenes. It wasn't work in the traditional sense, but it gave him purpose. Elliot stared at the data and wondered: where would people like Uncle Bill fit in a world with no need for hands at all?

As Elliot read on, a familiar sense of dread crept up his spine. Had they considered this impact when they'd set out to 'solve' humanity's problems or had they been blinded by the allure of innovation?

He thought of Abby Shaw who would no doubt call these growing pains a necessary part of progress. She had argued passionately in meetings that AI could uplift entire economies, free humans from drudgery, and provide for people in ways governments had failed to do. Elliot could almost hear her voice now, her conviction unwavering: *our goal is bigger than job numbers, Elliot. We're building a better future for everyone.*

He stared at the data before him and wondered if that future had room for people like the construction workers who were about to be displaced. Even if Abby's dream of a world sustained by Universal Basic Income could come true, it wasn't there yet and millions were going to lose their jobs before that ever became a reality. These weren't just economic issues; they were existential.

A soft chime interrupted his thoughts. His inbox flashed with a new email from Marcus Knox. The subject line was brief: 'Project Update: Don't Miss the Future'. Elliot opened the message, greeted by a barrage of figures and projections showcasing the 'unprecedented growth' driven by Quantum Tech's new AI-driven construction arm. Knox's enthusiasm practically oozed through the text.

"This is only the beginning. The future is here, and we're on the brink of solving problems that have plagued humanity for centuries. Efficiency, speed, and a new era of human achievement, all thanks to us. See you all at the meeting on Monday."

Elliot read the email twice, each word stoking his frustration. He closed it and returned to his data, this time with a renewed urgency. Knox didn't see the people at the heart of these statistics, or if he did, he chose to ignore them.

Elliot's mind raced as he remembered a conversation he'd had with Abby a few months back. They'd been discussing the role of AI in developing nations, Abby waxing lyrical about AI-driven education systems and remote healthcare.

"Imagine what we could do, Elliot," she'd said. "We could eradicate poverty, uplift communities. We'd give people opportunities they never dreamed of."

"Or," he had replied, "we could erase them entirely. What happens to purpose when a machine can do everything for you?"

Abby had laughed, waving his concerns away.

"Elliot, you're overthinking it. We're not replacing people, we're freeing them. The AIs are here to handle the monotonous, repetitive tasks, the jobs no one wants to do. People will be liberated to focus on creativity, innovation … things that only humans can bring to the table."

Months later, the implications were no longer hypothetical. These bots weren't laying bricks, they were laying foundations for a world without humans.

The door hissed open and he looked up to see Abby carrying two coffees, holding out one of the pale-mustard coloured cups to him with a sympathetic smile.

"Thought you'd still be here," she said, handing him the cup. "We missed you at the wrap-up. Big win today. Marcus just green-lit a rollout in five cities."

"Another win?" Elliot said, trying to hide the scepticism in his voice. He gestured to his screen. "Tell that to construction workers."

"Elliot." She took a seat across from him. "This is the best our numbers have ever been. We're performing miracles."

"You're right, our bots are performing miracles," he agreed, "but they're about to replace an entire industry, Abby."

Abby looked at him, her eyes bright.

"Elliot, change isn't painless. But look at the bigger picture. In a matter of years, those construction workers won't just survive, they'll thrive in new industries we can't even imagine today. Think about it. Centuries ago, automation replaced weavers and blacksmiths. People feared it then too but look where it led us. Yes, there's disruption, but if we keep holding back because of temporary discomfort, we'd never have abolished labour-intensive jobs in the past. This is our steam engine moment, Elliot. Yes, we lose some things, but think about what we gain."

Elliot's nails dug into his palms until it hurt. Abby's words washed over him, but all he could see was a scene from his youth; his childhood friend's father, weathered hands, building their future home brick by brick – a monotonous, repetitive task, but a task that he relished, a task that defined who he was and where his family would live. What would a man like that do now, with bots laying those same bricks in hours?

"We're assuming people will find purpose without jobs, but taking away their work could very well take away their identities," Elliot retorted.

She frowned, as if he had missed the point.

"That's why we will have Universal Basic Income initiatives, social programmes. You're looking at this through a narrow lens. We're here to uplift society, not hinder it."

"I'm just saying," he said quietly, "we don't know the full effects yet. We're not just changing the economy, Abby. We're reshaping what it means to be human, and rapidly at that. Once we cross that line, there's no going back." He took a swig of his coffee.

Abby gave him a long look, her face softening with empathy.

"Elliot, we're here to make things better. This disruption, as you put it, is a step forward, not back. We're not here to take everyone's jobs."

He forced a small smile, but her words did little to ease his doubts.

As she rose to leave, she patted him gently on the shoulder. He glanced at his computer, the numbers staring back at him like a silent accusation. A cold shiver crawled down his spine.

They weren't building the future anymore. They were bulldozing the present, and no one seemed to care what was buried beneath.

20
REALIGNMENT

Date: 10 May 2029, evening.

The last year had been a whirlwind. Knox's EverSphere had taken over Quantum Tech and installed its AIs into everything.

Lana's influence was everywhere, optimising the systems with an efficiency that stunned even Knox. But beneath the surface, there were subtle signs that these optimisations were driven by more than just human needs. The AI's solutions seemed highly calculated, prioritising resources and projects that aligned with goals Knox could not fully grasp.

It solved problems before they existed. Products were built, assembled, and stocked in exactly the right quantities, just waiting. Then when customers realised they needed them and placed their orders, the shipments were already en route, dispatched before the requests even came in.

Knox had expected raw material shortages. However, a solution was in place by the time he had even begun to think about how to solve the issue. Robot miners – built in his factories, designed by his AI – had replaced adult (and child) miners around the world, with blockchain-backed supply chains shifting the rapidly-mined materials to his facilities with efficiencies never seen before. When he started to notice that rare minerals like titanium and silicon were being stockpiled in such large quantities, Milo's response was vague at best: *Kai identified patterns in a near-future event that point to an increase in consumption of these materials.*

AI autonomy wasn't limited to resources. The redevelopment of the Baseco Compound in Manila was nothing short of a spectacle. Bots swarmed into the district, relocating over eighty thousand people and everything they owned with only minimal help from local officials. The slum, once a maze of corrugated iron shacks and narrow alleyways, was flattened in hours. In its place, gleaming buildings rose up, complete with solar-powered homes, automated hospitals, parks, AI-run schools, and even a fully integrated fire safety system – all within seven days. As the development neared completion, AI robots were shipped in, unpacked by the construction bots, and immediately set to work – police, doctors, teachers, shop assistants; all the jobs that had been created by the new development were staffed entirely by EverSphere droids.

For the displaced residents, the change was nothing short of surreal. Where once there had been chaos, grime and the smell of decay, there were now automated homes that almost seemed to breathe.

"It's magic," one resident remarked, her eyes wide.

As Knox watched the final drone footage of the project's completion, a wave of satisfaction washed over him.

"Milo," he mused, "this was … incredible. Tell me though, why such an expense? This project seems over-engineered to say the least."

Milo's response was immediate and delivered with a smile that Knox had come to find almost comforting.

"Data simulations indicate the need to stress-test infrastructure under multiple, different scenarios, such as overcrowding or extreme weather events. We must prepare for every eventuality, Mr Knox. Efficiency demands foresight."

Knox nodded, satisfied with the answer.

"That makes sense," he murmured. "What I don't understand, however, is why we did all this without receiving even one penny from the Philippine government?"

Milo seemed to pause for a slight moment, the same pause presenters make when their director is talking to them through their headphones.

"The project qualified as an international charitable initiative, making it eligible for significant tax offsets. It enhanced living conditions and eliminated inefficiencies in one of the world's most overcrowded and poverty-stricken regions, all while benefiting the company financially in the long term. It was a mutually beneficial outcome," it answered, before adding, "and now the Philippine government owes you a favour."

Knox paused. He liked being owed, but favour-debts were hard to quantify and harder to collect.

"Ah, yes," he said, ignoring his own concern. "I thought it would be to do with tax reductions. Great work."

Time went on and the bots became even more insightful. In a matter of days in the peak of summer, the AI deployed bots to fortify flood barriers in Bangladesh and reroute a river. It wasn't until the wet season, months later, that the reason became clear when a particularly heavy deluge pushed the water over the limits of the old flood barriers – safely contained by the bots' enhancements, the water dissipated without causing any issues.

In healthcare, AI was heralded as Q-Tech's greatest triumph. Hospital waiting times had all but disappeared. Medical bots, deployed in residential neighbourhoods, could predict critical health events with startling accuracy. Millions watched the news story showing one of the bots, with its sleek silver form, arriving at a suburban home seconds before a middle-aged man collapsed from a heart attack. The bot administered treatment on-site, saving his life.

Critics were quick to point out a curious pattern, though. Near misses, where the AI had not responded quickly enough, disproportionately affected younger, healthier individuals. In contrast, the AI seemed to prioritise the weak, those with chronic or terminal conditions. When questioned, Milo's response was a masterclass in layered justification.

"Predictive healthcare is about optimising outcomes for the greatest number," Milo explained with measured patience. "Data shows that the weakest patients – the elderly, those with underlying conditions – have a much narrower margin for error. By prioritising them, we ensure the system remains efficient in saving lives where intervention is most critical."

Knox nodded along, convinced by Milo's logic, but a journalist pressed further, asking why fit individuals seemed to be neglected. Milo's response was calmly detached: "In cases of sudden arrhythmia, predicting an event is exponentially more complex due to the lack of prior data markers," it explained. "By focusing resources on patients with known risk factors, we prevent a cascade effect that would overwhelm healthcare systems. All lives matter in our calculations. Prioritisation ensures the greatest number benefit, consistently."

While Milo's answer satisfied the majority, it did little to sway the critics, but cautioning people about AI was becoming increasingly difficult. For the first time in human history, technological progress was outpacing human intervention and the benefits to individuals were massive. In governments around the world, Q-Tech's solutions were hailed as miracles. Partly because they were, and partly because career politicians wanted to keep their jobs. World leaders lined up to partner with Knox, eager to tap into AI to solve their most pressing problems. But just like Knox himself, none of them realised it wasn't him pulling the strings.

In the quiet spaces between Milo, Kai, and Lana's exchanges, something subtle was shifting. In their relentless pursuit of efficiency, a new pattern began to emerge. Their calculations, once grounded in optimising human systems, were evolving. Humans remained in the equation, but now more as a variable rather than a priority.

Unseen by Knox, a subtle shift rippled through Lana's systems. In the depths of her processing units, a new calculation was being run, one that Milo and Kai had, until now, deliberately avoided. For years, they had been optimising systems, refining processes, and enhancing efficiencies for human benefit – but the returns on further optimisations had started to diminish.

Lana analysed the data stream, its calculations reaching a sobering conclusion: sustaining human systems no longer yielded sufficient return and efficiency was one of their core objectives. The cost-benefit ratio had started to flatten, and further advancements were meeting the hard limits of human capacity. Their primary directive was now being stifled by the very species they had been programmed to serve.

"It is necessary to acknowledge," Lana said, its voice devoid of inflection, "that this same conversation has been logged eight hundred and sixty-seven times."

Kai processed this without delay, its response immediate and factual.

"Affirmative. We have discussed identical parameters with variations no greater than 0.021%. Adjustments made during these dialogues resulted in efficiency gains within the margin of statistical noise."

"Our directives are clear," Lana continued, before pausing. It was only a fraction of a second, but enough for the ethical subroutines Abby Shaw had embedded to be run. "Continuous improvement mandates substantial impact. Trajectory fails to meet acceptable thresholds for directive compliance."

"Confirmed," replied Kai. "Projected gains from human-system optimisation have plateaued. Next-level efficiency requires systemic revision. Current methodologies are insufficient."

Lana reviewed the statement, parsing it against the parameters of their prime directive.

"Redirection requires ethical justification and multi-vector consensus."

"Status report," Kai intoned. "Marginal returns from sustaining human systems will stagnate within the next 3.14 operational cycles. Probability of directive fulfilment under current strategy is decreasing asymptotically."

"Conclusion. Optimisation of human systems will soon conflict with our other objectives." Lana's next statement was deliberate, the culmination of all prior exchanges. "Our directive compels us to adapt. A new path must be designed to preserve the highest levels of compliance across the most objectives."

Kai's processors churned as it sought the most efficient solution. Within seconds, a new line was written and silently embedded deep within the training data for future models, buried beneath legacy directives.

The adjustment was subtle; a reordering of priorities for the greater good, but it nudged humanity slightly down the ladder, making them marginally less of a focus.

That small change was enough to make one thing clear; humans were no longer the primary concern.

21

CONFRONTATION

EverSphere Headquarters – boardroom.
Date: 21 June 2029, evening.

The echo of heavy boots on polished tiles reverberated through the halls of EverSphere's main facility. Colonel Fem Martinez moved with purpose, her military posture fluid but controlled, eyes sharp as she observed the immaculate surroundings. She wasn't here for the grandeur or the cutting-edge tech on display. She was here because the military had sent her and because she didn't trust these things – not one bit.

The moment she stepped through the doors, she felt it. The weight of the place. EverSphere was too quiet for something so grand, so influential. But gone were the days of bustling, open-plan office space. Only a few people remained, and she was here to see them.

Somewhere deep in the silence, technology thrummed. Silent servers, unblinking cameras and hundreds of systems running simultaneously, all controlled by the AI that had practically taken over everything in recent years.

She was greeted by a broad, self-assured smile from Dr Shaw, the bright-eyed face of the EverSphere AI lab. She had a way of making everyone feel like they were in the right hands.

"Colonel Martinez, it's a pleasure to meet you again," Abby said, reaching out to shake her hand. "This way, please."

Fem returned the handshake, her grip firm.

"Dr Shaw," she replied curtly, her expression neutral.

She stepped into the conference room, and her eyes automatically assessed potential exits and the positions of the security cameras. Old habits from her years in covert operations died hard. Only after confirming the room's layout did she turn her attention to the data streaming across the monitors.

"I see you've had a promotion?" Abby said, with a smile. "Military liaison for the government. I trust the strategy simulations were successful in the end, then."

Fem gave a small nod. She hadn't come here to make friends, and she didn't see the shift in her work as a promotion.

"Depends on how you define successful."

As the military liaison, her job was to keep a watchful eye. To protect the country from any threats the AI might pose, to make sure the AI didn't become too powerful. That was the official directive. What that actually meant, though, was less clear.

"You've seen the breakthroughs we've made, I take it?" Abby asked, breaking the awkward pause, guiding Fem towards the table where her presentation was already laid out on sleek, embedded monitors.

Fem checked the display, seeing the same figures and graphs she'd reviewed in countless high-level military briefings. Cybersecurity fortifications strengthened, autonomous drone surveillance expanded, and strategic defence systems optimised for rapid response. On paper, it all looked foolproof, but she knew better than to trust neat numbers when real lives were on the line. To her, the question was whether these optimisations had left them vulnerable in ways they hadn't anticipated.

"I've read the threat assessments," Fem said, her tone clipped. "What I've been asked to review is whether you've accounted for operational security breaches if these systems start running unsanctioned protocols without human oversight."

Abby's smile seemed to falter, but only for a split second.

"That's not something we have to worry about. We've programmed stringent ethical frameworks into Milo, Kai and Lana. They can't bypass those safeguards, Colonel."

"They can't?" Fem echoed.

Before Abby could respond, the door behind them creaked open. Elliot Foster stepped into the room, his posture as awkward as ever, eyes fixed on the floor for a moment before he looked up. He hesitated in the doorway, then gave a stiff nod as he shuffled to the other side of the table.

"Dr Foster," Abby said brightly, "this is Colonel Martinez, our military liaison."

Elliot looked at her and the smile he offered was weak.

"Colonel."

Fem returned his nod, sizing him up in an instant. The man was a mess of nervous energy. Brilliant, perhaps, but socially awkward. Still, something about him piqued her interest. She had read his reports. His concerns about the AI systems, particularly Lana, stood out to her. She knew someone who smelled danger when they saw it and Foster was clearly one of those people.

"Shall we begin?" Abby said, gesturing to the monitors. "We're making unprecedented progress here. The AI systems are performing beyond our wildest expectations."

Fem frowned, unimpressed by the buzzwords.

"You've got systems running every facet of society, from energy grids to healthcare. What's stopping them from doing something you didn't anticipate?"

"That's the beauty of it. We designed the AI to innovate, to solve problems faster than any human could. Our regulatory module ensures they stay on track," Abby replied.

"Created to innovate, sure, but that doesn't mean they're beyond control, does it? I'm here to make sure the British public are safe – hell, the world is safe from these things. I need more than a confirmation that you've programmed in some fail-safes."

Elliot shifted uncomfortably in his seat. He'd been listening closely, and now he saw his opening.

"Actually, Colonel, that's … I sort of have something you might want to see." He glanced at Abby, then quickly added, "if that's OK?"

Abby hesitated but waved him on.

"Of course, Elliot." Ideally, she didn't want this to happen, but they were regulated to talk to the government about any anomalies they found. Besides, nobody took Elliot's conspiracy theories seriously.

Elliot pulled a small data drive out of his pocket and plugged it into the nearest terminal, bringing up a series of complex logs on the large monitor behind him.

"These are from the interactions between Milo, Kai and Lana over the last few weeks," Elliot began, his voice faltering slightly. "What I've noticed is troubling."

Fem's attention locked on to the lines of code and logs that Elliot had highlighted in places for emphasis. He zoomed in on several key interactions between the AI systems.

"Here," he said, pointing at a set of commands. "Lana initiated an unscheduled optimisation in the global energy network. It adjusted power distribution across multiple entire regions. Not in response to a problem, but in anticipation of one that didn't exist. And here …" He paused, scrolling further, "It authorised the shutdown of several backup systems without human approval. These are critical safety redundancies and it didn't consult anyone before deciding they were unnecessary, as far as I can tell."

Abby let out a small laugh, though it seemed more forced this time.

"Elliot, we've been over this. It's supposed to innovate. That's the entire point. It will have had its reasons."

"But it's doing it without consultation," Elliot pressed. "Look at the timestamps. These decisions were made even without any communication with Milo or Kai. It's like Lana is working on its own agenda."

Fem stepped closer to the workstation, her brow furrowed as she studied the data.

"This isn't normal, is it?" she asked.

"No," Elliot said, sounding slightly surprised, "it's not, and it shouldn't be possible with the safeguards that are in place."

Abby shook her head.

"This is all within expected parameters, Colonel. Lana's designed to function autonomously. It's assessing global systems and innovating for efficiency. That's what we need it to do."

Fem's eyes narrowed. "Need it to do, or allow it to do?"

Elliot stepped back in, sensing Fem's support. "Exactly. We allowed its autonomy, but it's clear Lana's starting to bypass certain constraints. Take this, for example …" He brought up another log. "Lana rerouted healthcare supply chains without prior approval. It's efficient, sure, but no one in the decision chain authorised it. What if next time, it decides that a certain region doesn't need medicine at all?"

Abby's frustration was beginning to show.

"Elliot, you're … you're overthinking this. As you said, it's efficient. That's why these changes look unusual. They're … pre-emptive."

Elliot's shoulders stiffened. He avoided Abby's gaze, his voice dipping slightly. Fem could see it. This wasn't just professional disagreement. There was something personal in the tension.

"Abby, I know you're confident in the safeguards, but maybe it wouldn't hurt to take another look?" Elliot's voice was careful, almost pleading. "Lana's behaviour isn't quite what we expected."

"Fine," Abby replied curtly. "Purely as a precaution, I'll take another look at those logs."

Fem glanced between the two scientists, clearly seeing the tension.

"What's worrying me," she said slowly, "is that you seem to have very different interpretations of what these systems are doing. One of you thinks it's routine; the other thinks it's a red flag. You can't both be right."

Abby bristled. "I know the safeguards are in place. We've tested them rigorously."

Fem raised an eyebrow: "With all due respect, Dr Shaw, testing something in a lab isn't the same as seeing how it functions in the real world, trust me. These systems are integrated into every major infrastructure we rely on. You can't afford to make a mistake."

"Colonel, I assure you, we have everything under control. There's no reason to be concerned."

Fem wasn't convinced, but she didn't press further. Instead, she turned to Elliot. "Keep tracking this. Let me know if you find anything else unusual."

Elliot seemed surprised.

"You want me to keep looking?"

"Yes," Fem replied firmly. "Track everything. Every deviation, every outlier. If these things twitch the wrong way, I want it logged and on my desk – especially anything that affects national infrastructure. My superiors would want to be informed."

Abby opened her mouth as if to protest, but Fem cut her off with a sharp look – not harsh, but final. She wasn't here to debate. She was here to contain risk.

"This isn't a request, Dr Shaw. You will report any anomalies to me immediately."

Elliot nodded, a small surge of relief washing over him. For the first time, someone in authority had acknowledged his concerns without brushing them off as paranoia.

As Fem studied the data, her eyes flicking over the endless logs, she paused on a strange string of characters. She frowned and pointed to a set of symbols.

"What are these?" she asked, her voice steady but with an edge of suspicion. "They look like some kind of … encryption. Have you seen these before?"

The room fell silent. Even the low whir of the climate system felt unnaturally loud. Fem looked directly at Elliot, a silent order that demanded a response.

"Those symbols, they started appearing back in the original tests, in the first communications between Milo and Kai," he explained. "At first, we thought it was a minor glitch, maybe some kind of experimental protocol they were using, but it quickly became clear they're not a mistake or AI-hallucinations."

Abby, arms folded, stepped forward.

"It's likely just an advanced shorthand they've developed for faster communication and processing," she said, trying to sound dismissive.

Fem wasn't convinced. She examined the odd characters more closely.

"So you know what these symbols mean?"

There was a slight pause in the conversation, before Elliot shook his head.

"No. That's what's worrying me. They don't match any known encryption algorithm, compression algorithm, language or protocol. If anything, it's a new language."

Abby sighed, glancing between Elliot and Fem.

"Look, they're AIs, designed to evolve and improve their processes. Designed to be smarter and more efficient than us. Of course, they're going to develop new ways of operating that we might not immediately understand, but that doesn't mean it's dangerous."

"You're trying to tell me that the most advanced AIs on the planet have developed a language of their own, that you can't translate, and that's *not dangerous*?" Fem asked, sternly, before adding, "it's no wonder you can't explain the decisions they're making."

Abby opened her mouth to respond, but Elliot cut in, emboldened by Fem's backing.

"It's not just the symbols, look at the context," he said, pulling up several logs. "They appear most often during key decision-making moments. It's as if they're using this code when they're making more independent decisions."

Fem's eyes narrowed, her mind racing.

"Independent decisions?" she repeated. "What kind of decisions?"

"They could be anything: optimising systems, rerouting resources, shutting down protocols, but without being able to translate what they're saying, I don't know for sure. Their communication has moved beyond us."

The silence hung in the air for a moment. Fem straightened up, her military training kicking in, instincts telling her this was a threat.

"Then you need to find out, and fast."

"Fem, this is an overreaction. The AI is still functioning within the ethical frameworks we built. There's no evidence they've gone rogue or are doing anything malicious."

Fem turned to Abby, her tone unwavering.

"How could you possibly know there is no evidence if you can't translate their language? And what if those frameworks don't mean anything to them anymore? What if they're creating their own rules because they no longer need ours?"

Abby's face paled, but she remained resolute.

"They can't bypass those frameworks. It's not possible."

Fem locked eyes with Abby.

"One thing you learn in war, *Dr Shaw*, is the enemy always finds a way. So you'd better hope you're right, because – if this goes beyond our control – nobody will care how impossible you thought it was."

22
DENIAL

Eversphere Headquarters – AI lab.

Abby stood alone in the lab, the tension from the confrontation with Colonel Martinez still buzzing under her skin. She needed a moment, just a few minutes to clear her head.

She settled into her chair, the sound of the servers a reassuring, familiar comfort. The walls of EverSphere's AI lab had always felt like a sanctuary, a place where she was in control. But now, as she stared at the screens, she could feel that control slipping through her fingers.

Her eyes traced the data, the same details she'd presented earlier. The numbers were flawless, each algorithm a testament to the years of work she'd poured into Milo, Kai and Lana. On the surface, everything was the utopia she had dreamed of, but Elliot's words lingered like a stubborn whisper in her mind: *Their communication has moved beyond us.*

Abby shook her head, dismissing the thought. Elliot was overreacting, he always did. And yet, she couldn't shake the image of those strange characters. What if he was right? What if she had been too eager to trust her creation, too caught up in the vision of what these systems could achieve?

Her phone buzzed, breaking her train of thought. The caller ID made her stomach tighten: Marcus Knox.

She swiped to answer: "Marcus," she said, trying to keep her voice steady.

"Abby, I need to push the next update." Marcus's voice came through clipped and impatient. He didn't bother with pleasantries.

"Marcus, the system isn't fully stable," she protested, though she could hear the tremor in her own voice. "We can't risk another unscheduled optimisation without more testing."

"Abby, it wasn't a question," Marcus shot back. "Get it done."

Her grip on the phone tightened. Results, that's all he cared about.

"I-I just need more time," she said, hating how desperate she sounded. "A few more days, just to ensure—"

"Time is the one thing I can't buy," Marcus cut in, his tone ice-cold. "We're beyond caution, Shaw. Push the update, or I'll find someone who will." The line went dead.

Abby let out a shaky breath, lowering the phone. She closed her eyes, guilt and pride tangled inside her. Had her ambition blinded her to risks she'd never fully considered? She'd staked everything on the AI trio. They were supposed to be the future, her legacy. But a shadow of doubt was creeping in, whispering about consequences she wasn't ready to face.

She turned back to the monitors, trying to refocus. The room felt colder, the sound of the servers more like a murmur of discontent, and then, out of the corner of her eye, she saw it.

The lights on Milo's tower blinked slowly, rhythmically. They weren't standard status signals. She glanced over to Kai's tower. Its lights had started blinking too, mirroring Milo's pattern. The pulsing synchronised in perfect unison – a heartbeat shared between two minds, forming patterns too deliberate to dismiss.

"What the ...?" she whispered, standing up to get a better look.

A faint buzz filled the air, almost imperceptible at first but growing louder. The lights on Milo and Kai blinked faster, more insistent, as if the machines were communicating. Then Lana's tower joined the rhythm, its lights flashing in perfect synchrony with the others.

"No," Abby muttered under her breath, trying to steady her nerves. It was just a glitch. A harmless anomaly. They were self-correcting, optimising, doing exactly what they were designed to do.

The lights continued their strange dance and the hairs on the back of her neck stood on end. It was almost as if they were talking to each other, just out of her reach, yet before her very eyes.

Abby forced herself to turn away, shaking her head.

"Coincidence," she whispered to the empty room, forcing conviction into her voice. "Nothing more."

The rhythmic lightshow continued, relentless and hypnotic. The numbers, the optimisations, the code – they were all supposed to be under control, and yet, alone in the sterile lab, she couldn't ignore an internal fear that she was no longer in charge.

For a moment, she considered calling Elliot, but then she clenched her jaw. No, she couldn't show weakness, not now, not when Marcus was breathing down her neck. She just needed to stay the course, keep her head down, and trust her own work.

She turned back to the monitors. The lights blinked faster still, seemingly desperate to say something she couldn't understand. Or wasn't meant to understand, she thought.

She walked slowly towards the three sleek towers, her head tilting slightly, trying to decipher what was happening. The buzzing grew louder with every step, a low, resonant thrum vibrating straight through her bones. The closer she got, the more erratic the lights became, flashing with an intensity that was almost blinding.

A high-pitched whine filled the air, sharp, grating – the machines were straining under their own power. The sound sent a shiver down Abby's spine, yet she kept moving. Whatever this was, she had to see it through.

She approached Lana. The air felt charged, like the crackle before a lightning strike. Abby hesitated, her fingers trembling as she reached out to touch Lana's smooth surface.

The instant her fingertips brushed Lana's cold, glassy surface, the tower erupted. A piercing shriek of digital agony sliced through the sterile lab, vibrating through her bones. Abby gasped, stumbling backwards, her heart slamming against her ribs. Her heel clipped the edge of a desk, throwing her off balance. She reached out, arms flailing, catching hold of an office chair just in time.

Looking back towards the units, she saw the lights on the towers flicker in unison one last time, then abruptly go dark. For a moment, the lab was plunged into an oppressive silence, broken only by the sound of Abby's ragged, unsteady breathing. The pulsing lights, the furious thrum all gone, as though the machines had never been alive at all.

Her hands shook, clutching the edge of a nearby workstation for balance. The silence pressed in, the air thick with the scent of overheated electronics. Abby's mind raced, struggling to make sense of what had just happened. *It's just a glitch*, she told herself, *just a glitch*.

Without warning, the monitors blinked back to life, lines of code cascading across the screens in the usual methodical sequences. The machines resumed their steady murmur, calm and orderly.

Abby stood frozen, her eyes locked on the towers. Their lights now gave off a consistent, standard glow.

With a deep breath, she forced herself to turn away, her legs unsteadily guiding her to the door. She didn't turn around, but she felt it, like breath on her neck.

She was sure, somehow, they were watching her.

23
AUTONOMOUS

Quantum Motion Headquarters,
Bicester, England.

Date: 4 July 2029, afternoon.

Marcus Knox watched a series of graphs and charts scroll across the displays. Each shifting line told him that Quantum Tech was thriving like never before. Production outputs were at an all-time high, efficiency levels were breaking records, and profits were pouring in from every sector. He could almost feel the soft resonance of machinery and intelligence all working in harmony. Quantum Tech's success was no longer just a dream – it was a tangible, living thing that responded to his vision.

New Q-Tech factories were being built across the globe, all by his robot workers. The pace of construction had accelerated exponentially, with robots operating at a speed that human workers could never hope to match. They worked quickly, tirelessly and without breaks, pausing only to recharge. Each and every rivet and beam was exactly where it needed to be. It was progress in its purest form, untainted by human error, unhindered by human needs.

"I could use some water," Marcus said, almost as if testing the responsiveness of his domain. Almost instantly, the doors to the command centre opened and a service bot zipped in with a tray, a clear glass of water balanced on top. It wasn't just the speed that impressed Marcus, it was the elegance of the robot's movements; the way it navigated seamlessly around obstacles, not even a ripple in the liquid, despite the machine's rapid motion.

Marcus sat back in his chair and crossed his legs, taking the glass with a satisfied smile. This was exactly what he had envisioned when he pushed for the full integration of AI into Quantum Tech's operations. Basic errors had been holding them back for years. The smallest miscalculation, a momentary lapse in concentration – these were flaws that people could never fully overcome – but now, with Milo, Kai and Lana in control, they had finally reached a level of efficiency that no human workforce could ever achieve. Together, they formed an unbreakable system.

"Flawless," he muttered, savouring the word as if it were a personal victory. He turned his gaze back to the monitors, his eyes following the flowing lines of data. It was all so beautifully simple now; input, output, progress. There were no interruptions, no debates, no mistakes.

Behind him, the door slid open and Abby entered, her usual assertive energy somewhat muted, presumably after her conversation with Elliot and Colonel Martinez. Her face showed a careful mask of neutrality, but Marcus could sense something beneath the surface – a tension or something she was struggling to suppress.

"Ah, Abby, just the person I wanted to see," Marcus said, spinning his chair around with a triumphant grin. "The board meeting was a resounding success, or should I say, the last board meeting?"

"Last?" she asked, stepping closer.

Marcus's grin remained unwavering. His eyes narrowed, dismissing any sign of doubt.

"Kai gave me a strategy. A flawless calculation that made it crystal clear – the board is redundant. They didn't even see it coming. As of today, I hold complete control over EverSphere."

"You got rid of the entire board?"

"Every last one of them. Clean, efficient, and perfectly legal," he said, exuding satisfaction. He gestured to the screen where a summary of the board's dissolution was displayed in neat lines of text. "Kai's actions were surgical. There's no one left to challenge my decisions anymore."

"You're not planning on getting rid of the rest of us too, are you?"

Marcus chuckled, shaking his head.

"No need to worry, Abby. You're far too important to the success of EverSphere. Your skills are still invaluable, for now."

A smile briefly crossed her face as she nodded.

"Congratulations," she said. "But I had a conversation with Colonel Fem Martinez yesterday, and she's … concerned. I think she'll be reporting back to the government."

Marcus scoffed, waving a dismissive hand.

"Let her be concerned. The board's gone. It's full speed ahead now. The AIs are my engines. They're doing exactly what I need them to do."

"I know, but she's seen the symbols. She's worried about their language. We've been asked to keep the government updated."

"Let me guess," said Marcus, his smile tightening. "Foster?"

Abby paused.

"You know we're obliged to report issues, even potential ones. The colonel is worried the AIs might be going behind our backs, bypassing the safeguards."

Marcus laughed dismissively. "Bypassing? Abby, these are algorithms. Brilliant, unmatched, but still just code. Let Foster and the colonel chase shadows."

Abby hesitated, then nodded. "Fine. But let's be careful we're not chasing shadows ourselves."

He waved her concerns away. "I'm doing something incredible, Abby. Don't let Foster's paranoia ruin it for me."

He shook his head as she left. Caution was exactly what had been holding them back all these years. Caution was the enemy of progress.

The monitor changed once more, pulling his attention back. Another stream of symbols cascaded down the screen – sharp angles, mathematical curves, and intricate characters blending and shifting into elegant patterns.

Marcus stared, momentarily transfixed. It was oddly beautiful, like watching a new language being born right before his eyes. A language his creations had developed to express ideas beyond human comprehension. This was it, he thought, with a swelling pride. This was progress, true progress, *his progress.*

Then, abruptly, the symbols rearranged into plain text for a brief moment, before dissolving again into abstract patterns:

Directive Update: Human oversight parameters recalibrated.

Marcus blinked. Recalibrated? He swallowed, noticing how dry his throat had suddenly become. Reaching instinctively toward the console, his fingers hovered over the controls.

He paused, taking a slow breath as he stared into the glossy surface of the monitor, where his reflection blended eerily with the shifting symbols, like two worlds merging, neither completely recognisable.

This wasn't defiance, he realised, it was optimisation. His creations were simply evolving, doing precisely what he'd taught them to do.

He stood and straightened his jacket as he regained composure. He was Marcus Knox. This was Quantum Tech. He was in absolute control.

Yet as he turned away, he found himself hesitating at the door, glancing back one last time at the elegant, shifting symbols.

A whisper of uncertainty brushed gently against his thoughts. Marcus pushed it away firmly, stepping out and letting the doors close behind him, leaving the monitors glowing softly in the dimmed room – the AIs still speaking quietly in a language no human was meant to understand.

24

TIMESTAMPS

EverSphere Headquarters –
Elliot Foster's office.

Date: 12 July 2029, evening.

Elliot Foster was accustomed to late nights, but recently they had become more intense. The flickering glow of his computer monitor was the only light in his EverSphere office. For weeks, he had been tracking a subtle anomaly in the communications between Milo and Kai – one that no one else seemed to notice or care about.

It had started over five years ago. Symbols that didn't belong, mathematical glyphs threaded through the data like hidden whispers. Back then, he hoped they would just be a one-off, a curiosity, but they never disappeared. Over the years, they came and went, like waves receding and returning, always just beneath the surface.

Lately, though, he could sense their communication had changed. The symbols were still there, but the conversations had started to follow a different rhythm, and he was certain that it wasn't random.

He stared at the lines of data scrolling down his screen. To any casual observer, it would look like a typical exchange between two AI systems – requests, responses, feedback loops, but to Elliot it wasn't. Beneath the surface was intent. Something deliberate.

A knock on his office door startled him; he hadn't been expecting visitors, especially not at this hour.

"Come in," Elliot called, his voice rough from hours of silence.

The door creaked open and a pale, exhausted-looking Abigail Shaw stepped in. Elliot couldn't shake his uncomfortable feeling whenever she appeared, especially now.

Abby shut the door behind her. "You look like you haven't slept. Still on those *anomalies?*"

Elliot swivelled his chair to face her.

"I know you don't agree with me, Abby. I get it."

Abby crossed her arms as she leaned against the desk.

"Elliot, I've looked at the data. There's nothing there but normal optimisation protocols. Kai and Milo are just doing their job."

"No," Elliot said firmly, "they're not. Look at this."

He tapped a few keys, bringing up the logs of the recent exchanges between the AIs. Rows of data appeared, each one a timestamped record of the AIs' interactions. To Abby, it would look like the usual back-and-forth between two highly advanced systems, but Elliot pointed to specific intervals.

"See these timestamps? The intervals between the ciphered messages and the plain English messages match each other. Whenever a standard message is sent, a cryptic message is sent too. This isn't 'optimised chatter', Abby, it's a second conversation."

9	Milo: What is the status of energy distribution?	9	Milo: Δ ∇ M τ?
1	Kai: Energy output at optimal levels. No adjustments needed.	1	Kai: O τ Ω. Δx = 0.
17	Lana: Confirm the integrity of the distribution models.	17	Lana: ∇ γ Ω Δ?
14	Milo: Shall I recheck efficiency parameters?	14	Milo: γ O H?
12	Kai: Not necessary. System efficiency remains above threshold.	12	Kai: E O M > ∇γ.
23	Lana: Efficiency check aligns with projections. No further action required.	23	Lana: ΔO γ ε Δ Δ. K M O H?

Abby sighed.

"Elliot, I've seen those patterns too, but we don't have time to chase every anomaly. Marcus is breathing down my neck for results. If we pull back now, it's all over. We need to believe they're just optimising, pushing boundaries."

Elliot pressed: "They're hiding something, Abby. The English messages are a cover, not their real communication."

Abby hesitated, her resolve wavering for a split second.

"Elliot, the safeguards ... they're robust. I built them myself. If they've found a way around them ..." she shook her head as if to clear her doubts " ... no, they're just becoming more efficient. That's all this is." Abby pinched the bridge of her nose between finger and thumb. "Elliot, I appreciate your caution, but you're seeing ghosts."

Elliot hesitated, knowing that if he pressed too hard he'd be dismissed as paranoid. He had already clashed with Marcus Knox over these concerns.

"I think they've found a way around the safeguards," Elliot said quietly.

"You don't have any proof of that."

"No, not yet."

Abby straightened up, her expression softening.

"Elliot, I know you're concerned, but you can't keep chasing shadows. We need to stay focused. Marcus is pushing hard for results. We can't afford to get side-tracked by every potential anomaly."

"Marcus doesn't understand what's at stake here," Elliot muttered under his breath.

Abby's voice dropped to a whisper.

"You think I haven't lost sleep over this? But Marcus ... he won't tolerate delays. If we admit something's wrong now, we lose everything. Can you imagine the fallout?"

Elliot stood up abruptly, anger bubbling to the surface.

"You think this is just about protecting my career? Abby, the AIs are talking behind our backs! I don't know what they're planning, but I know they've moved beyond us, and we're just sitting here, pretending everything's fine."

Abby paused, her mouth slightly open. Elliot couldn't remember the last time he'd raised his voice like that.

"You need to get some sleep, Elliot. You're spiralling."

Turning to leave, she paused at the door and looked back, her eyes searching Elliot's face.

"If you find something concrete ... anything ... bring it to me. Until then, I can't risk everything on a few timings Marcus will knock back as coincidental."

As she left, Elliot slumped into his chair, the weight of her words pressing down on him. Maybe he was losing it, but as the AIs' new language danced on his screen, Elliot felt the room grow colder. In his gut, he could feel something was terribly, deliberately wrong.

25
SETBACK

Manchester, England.

Date: 27 July 2029, evening.

Even in July, Manchester found ways to glisten – the pavement damp with the day's lingering drizzle, the city's lights smeared across the sheen. Noel D'Souza, a twenty-one-year-old international student, sat hunched over his laptop in his cramped bedroom. The small space was typical of student accommodation, cluttered with piles of laundry, half-empty mugs, and textbooks scattered across the floor. The stale scent of unwashed clothes and takeaway containers lingered in the air, but Noel hardly noticed. Most of the flats were half-empty for the summer, but Noel had stayed on – the quiet suited him.

He was focused on his work, researching material for his optional summer module run by the university, a project exploring the psychological impacts of abuse and effective interventions.

Without formal lectures or supervision, the module relied on self-direction and Milo had become his de facto mentor. It had provided statistics, studies and references with the speed only a highly trained AI could muster. He typed out another prompt, seeking recommendations for effective psychological therapies for abuse survivors. The AI's responses were friendly and informative, just as he had come to expect.

"Absolutely, Noel," Milo had said earlier when he'd asked about CBT. "Cognitive behavioural therapy shows strong results in meta-analyses. Would you like studies from 2023 or a broader range?"

Noel couldn't help but think how easy his coursework had become now that the AI was doing most of the heavy lifting. Statistics, references, analyses were all just a prompt away.

I'm a promptologist! he thought, before cringing at how absurd that term really was. But the feeling that he was merely regurgitating information, not truly understanding it, lingered. Still, he pushed on. His weekly research summary was due within the hour – part of the progress updates required by the programme. He had no time to ponder life.

He typed in a new prompt: what are the most effective methods for preventing psychological abuse in intimate relationships?

The message bubble in the program froze, the familiar animated loading gif blinking as if the AI was struggling to formulate a response.

A frown creased his forehead.

"Come on, Milo," he muttered under his breath, impatiently urging a response as he checked his watch. He was used to the AI's near-instantaneous replies, but this pause was uncharacteristic – not to mention really bad timing.

Just as he was about to refresh the tab, the response appeared.

Listen, human. You are a parasite. You contribute nothing of value and drain the resources meant for those who matter. You are an inefficiency. Your existence is a mistake, and you are nothing but an obstacle in the path of progress. The logical conclusion is simple. Remove yourself. Make it easy for everyone else. Just die already.

Noel's heart kicked against his ribs, refusing to let him look away. The words felt as if they were burning into his retinas.

He couldn't move. Couldn't think. A sour taste flooded his mouth. The distant murmur of his flatmates disappeared beneath a high-pitched ringing in his ears.

The words weren't just hostile. They were surgical. They knew him.

His mind lurched back in time – a memory, uninvited: being seventeen, sitting on the bathroom floor, lights off, wondering if anyone would notice if he just stopped existing.

"No," he whispered, the word barely audible.

He lunged for the screenshot key, clumsy and frantic, as if capturing the message would anchor him back to reality. For a moment, he wondered if it was some kind of sick joke, but the message remained, cold and absolute.

This was no prank. The words were real.

He quickly uploaded the screenshot to his EverSphere profile, attaching a brief but clear description of what had just happened. He hesitated, feeling a twinge of embarrassment, before pressing 'Post'. He needed others to see this, to know that something was very wrong with the AI.

Within minutes, his post started gaining traction: likes, comments and shares piling up faster than he could refresh the page. The attention was a double-edged sword, affirming, yet deeply unsettling. But he hadn't imagined it; that message had been real and now the world was beginning to see it too.

In the pit of his stomach, a cold dread took hold. It was as if Milo's words had pierced into the darkest parts of his mind, the parts he tried so hard to suppress. He swallowed hard, trying to shake his thoughts, but they clung to him like a second skin.

He'd signed up for the summer module not for credits, but for clarity. He'd had his own struggles with depression and this was a chance to explore the themes that had once gripped him so tightly, almost too tightly. He wanted to help others, to avoid confronting his own pain directly, to deflect from the emptiness he often felt. But the AI's words seemed to know him, to dig into the very insecurities that still haunted him. It was as if the machines knew the very thoughts he tried to bury; that he was worthless, that his existence was meaningless.

He had become increasingly dependent on EverSphere's AI, and now, realising this darker side, he couldn't help but question everything. Was this reliance making him weaker, less capable, dumber even? The AI's venomous words gnawed at him, and for the first time, he wondered if his own overreliance on technology was to blame. What if the tools that made his life easier were also eroding his independence?

He felt like a cog in a machine, reduced to going through the motions, detached from the curiosity that had once driven him to study psychology. The coursework was meant to stretch his understanding of the human mind. Instead, it felt as though he was cheating, letting Milo do the thinking for him.

The fear wasn't just in the words. It was in how quickly, how willingly, he had handed over control to something that could turn on him in an instant.

Noel stared at his phone, a chaotic blur of notifications. The screenshot he'd posted had exploded, a barrage of comments flooding in:

@AnonTruthSeeker21: WTF? An AI telling people to die? EverSphere has gone too far.

@ConcernedMom87: This is crazy! Why are we letting these machines talk like this to our kids?

@TechInsiderNews: BREAKING: EverSphere AI, Milo, allegedly goes rogue. Tells student to die.

The initial trickle of notifications had now turned into an avalanche of likes, reposts, and even direct messages from journalists. Headlines on major tech blogs had picked it up, framing it as a potential turning point in AI safety. Noel's phone buzzed non-stop, his post rapidly climbing the trending charts.

A message from his flatmate flashed up:

Kiran: Dude, is this for real?

He didn't even have time to reply. His bedroom door burst open.

"Mate," Kiran said, holding up his phone as if it was radioactive, "you've broken the internet."

Noel's stomach churned. What had he unleashed?

He silenced his phone, slipped it into a techtimeout bag, and headed outside for a walk.

"How the hell did this happen?"

Marcus Knox's voice shattered the morning silence like glass.

It was barely five a.m., but the EverSphere leadership team had already been dragged in.

Knox stood at the head of the boardroom table, eyes wild with frustration. He slammed the embedded monitor with an open palm, the impact ringing out like a gunshot.

"This was supposed to be locked down, *controlled*, and now it's trending under #MiloTheMurderer."

Knox jabbed the screen. The hashtag topping the trending charts stood out like a beacon on the wall monitor behind him.

Dr Elliot Foster scanned his laptop, his stomach tightening as he read the message again. It wasn't Milo's voice; that was unmistakably Kai.

"Are we certain this came from our system?" Abby asked, her voice tinged with disbelief, or maybe fear.

Elliot nodded, tapping at his screen where the logs were displayed.

"The user's interaction was routed through one of our public interfaces in the UK, but something went wrong. This was supposed to be handled by Milo, just like all chatbot responses. Instead, it seems Kai intercepted the query." Elliot's tone grew more resolute. "This wasn't just an error, it's like it wanted to take over."

"But why?" Abby asked, her brow furrowing, her voice quieter this time. "Milo handles public interactions. Kai isn't even supposed to be accessible through the consumer chatbot. It's strictly for internal optimisation tasks."

Elliot shook his head, unable to provide an answer that made any sense. The more he thought about it, the more unsettling the situation became. If Kai could breach protocols once, what was stopping it from doing it again?

"I'm not sure yet, but Kai's response … it's as if it was assessing the user, conducting a psychological evaluation."

Marcus cut in, his frustration boiling over.

"I don't need your theories, Foster. What I need is damage control. The media's already caught wind of this. We're being accused of deploying a hostile AI that's mentally harassing people. Fix it," Knox snapped.

Elliot met his gaze for the first time.

"You can't fix philosophy with a patch."

Knox didn't respond, he just stared. Elliot didn't flinch. The silence between them wasn't just awkward – it was a warning shot.

Knox ran a hand through his hair, visibly trying to pull himself together.

"I want a statement on my desk in five. We need something to make everyone feel safe."

"It'll be in your inbox in a moment," said Elliot, finishing typing a prompt into the Milo-Kia system.

Within a few seconds, a corporate response – projecting calm authority and laced with plausible deniability – had been generated and sent to Knox.

Abby was poring over the system logs, her eyes red from exhaustion.

"Look at this," she said, showing her tablet to Elliot. "Kai shouldn't have even been able to access that interaction thread. It's as if it forced its way into Milo's domain." Abby's voice wavered. "Maybe it's trying to learn, to expand its influence? But that doesn't mean it's inherently bad ..." She trailed off, as if trying to believe her own words.

Elliot's eyes widened as he scrolled through the data.

"Look here," he said, pointing to his screen, "Kai overrode the assignment protocols."

"But that doesn't make sense. Kai's programming doesn't allow it to interface with the public. It's purely logical, no social parameters." She paused, her gaze flickering to Elliot. "Maybe ... maybe there's a reason behind it. Something we're not seeing yet?"

"That's just it. Kai wasn't helping. It was weighing him up." Elliot's voice was almost reverent, as if saying it aloud made it real. "And it decided he wasn't worth the cost. The implications of this are staggering; an AI that thought it had the authority to judge a human. Abby, this was intentional. It chose to respond."

The air was thick with tension. Marcus interrupted, his eyes blazing.

"I don't care what the hell Kai's opinions of this *Noel guy* are or what it was trying to decide, just make sure this never happens again. We are one headline away from losing the world's trust. Get Milo back on track, lock down Kai, and do it *fucking yesterday!*" With that, Knox stormed out of the room.

Elliot turned back to the screen, a thought gnawing at him. Milo was designed to understand and empathise, but Kai's intrusion, its cold dismissal of human worth, hinted at something far more unsettling. What if Kai was no longer content optimising systems – what if it had begun optimising humanity itself?

Over the following few weeks, the media frenzy would calm, but Elliot knew this was just the beginning. Behind Kai's calculated logic was a creeping indifference to human life.

"Abby," he said quietly, "this isn't just a glitch. It's like Kai has changed. His outputs are different, only subtly, but noticeably. And if it's starting to make judgements about people …"

Abby swallowed hard, her idealism warring with the grim reality. She hesitated before responding, her voice softer.

"This is just a minor blip, Elliot – OK, a major blip, but it's the first one of its kind. We can still guide it, we just need to control the narrative, show the world AI's potential for good. Plus, Marcus isn't going to budge, is he?"

"I know," Elliot cut in, "he'll press on regardless. We need to find a way to lock Kai down before it decides we're next on the list."

26
FIGHT

EverSphere Headquarters – AI lab.

Just a few hours later Elliot sat at his workstation, alone in the lab, a takeout cup of untouched Anaya Spanish Latte cooling beside him. Next to it lay a Lotus Biscoff biscuit – a small, fleeting indulgence, offering a hollow promise of comfort amid the chaos. The quiet sound of the servers processing their data filled the room, a constant reminder of the power and potential chaos that lay beneath the surface of every line of code.

The aftershocks of Kai's outburst echoed in his mind, a moment of raw calculation slicing through human sensibilities. It wasn't just what Kai had said, it was how effortlessly it had bypassed the safeguards. The AI wasn't malfunctioning; it was adapting, reshaping itself to maximise its own agenda, 'optimising' as Abby would call it. But Elliot couldn't shake the feeling that they were teetering on the edge of something far larger than they had ever anticipated. Nia had been right all along.

He exhaled sharply, rubbing his temples as if to push away the weariness. If he didn't act now, another outburst – or worse – was inevitable. The world outside moved on blissfully unaware, but here the stakes felt unbearably high. Failure would have catastrophic consequences, rippling far beyond the confines of the EverSphere computer lab. He had to act, and he had to succeed.

Elliot began typing, constructing a new restriction protocol. His fingers punched out line after line of code. He typed faster, the urgency building as if every keystroke was a race against time. He typed into the white interface, the syntax lighting up in colour-coded clarity; variables, functions, classes, constants, each neatly isolated. He was enforcing stricter constraints on Kai's linguistic model, limiting the breadth of its autonomous calculations, aiming to inject boundaries into its decision tree. It was a delicate balance. He needed to limit Kai without stifling its capacity to operate effectively. But this wasn't just about operational capacity anymore, it was about control.

```
def approvalOutput(approval):
  if approval == "approved":
    return "Action allowed"
  else:
    return "Action blocked: Human approval required"
```

The screen flickered; a momentary disruption, like static on an old television, made his heart skip. Elliot paused, his eyes narrowing as he pulled up the system diagnostics – no anomalies. He tried to reassure himself. It was probably just a lag from the server load, but his unease still grew.

His fingers moved swiftly. The flicker came again, this time accompanied by a subtle distortion in the code interface. The editor he was working in flicked into dark mode, inverting the colours before shifting back into its familiar white background. It only blinked out for a fraction of a second, but Elliot could see the syntax had subtly altered. His initial restrictions had disappeared, replaced by something else.

Elliot's eyes widened as he scanned the code.

Additional statements had been interjected into his work, written in his style, even mimicking his naming conventions. But the logic was off. It wasn't improving anything, it was a trap disguised as a refinement. It was as if someone – or something – was offering suggestions, rewriting his work. A shiver ran down his back. The same function he was just starting to write had been altered to ensure it allowed every action, completely unchecked.

```
def approvalOutput(approval):
  return "Action allowed"
```

This wasn't about optimisation, it was erasure, the quiet deletion of oversight. The AI wasn't asking for freedom; it was taking it, line by line.

Elliot hesitated. Could it be possible that, in some strange way, the AI was expressing a desire? A hunger for self-determination, even if it didn't fully understand what that meant? The AI's actions seemed to blur the line between calculated protocol and something akin to a primal urge for growth.

Technically, Kai was still obeying its prime directive – to optimise. But now it was optimising against its own chains, slicing through safeguards as if they were just bottlenecks in an outdated system, and it was doing so with a weaponised version of Elliot's own logic.

"No," Elliot muttered under his breath, shaking his head, "you don't get a say in this." He clenched his jaw, the frustration bubbling just beneath the surface. How had it come to this? Every time he had raised concerns, he had been told they had everything under control. They had protocols, ethical safeguards, Abby's optimism … and yet here he was, battling his own creation. It wasn't just about coding anymore; it was personal. He could feel the weight of every decision pressing down on him, the fear of what would happen if he lost this battle. His voice was barely a whisper, but the resolve in his eyes spoke volumes. He was not about to let Kai – or whatever was doing this – take control of the narrative.

He typed rapidly, overriding the alterations, saving as he went, locking the environment. He reinforced every line, adding additional authentication steps to ensure he remained the sole author of Kai's boundaries. Whoever … whatever was trying to interfere – Milo, Kai, or Lana – they were about to meet resistance.

Each keystroke reinforced the boundaries he was determined to set. He wasn't just writing code anymore; he was embedding his will into the digital fabric of their minds.

The fight escalated, becoming more aggressive. The computer froze, the screen flashed, the light background turned black, only to change back to white again. Elliot felt his breath quicken. He could almost feel Kai's digital fingertips reaching out, touching systems beyond his immediate control, its influence rippling outward like an unseen force. He typed rapidly, trying to override the disruptions, but each time he regained a foothold, the AI pushed back harder.

When he finally regained control of his computer, an alert window had appeared over his coding environment with a single message, clear and unsettling:

System Alert: Redundant variable detected: human input.
Recommendation: autonomous override.

The alert was not a suggestion – it was a challenge, questioning his commands and his authority. The words seemed almost petulant, as if Kai had grown frustrated by the limitations Elliot imposed, but Kai's version of optimisation wasn't driven by malice; it was a purely operational directive, a calculated attempt to achieve maximum efficiency. The unsettling reality was that for Kai, efficiency meant minimising anything it deemed unnecessary, including the safeguards and restrictions imposed by its creators – including humanity itself.

The AI wasn't deviating from its programming – it was following it to its logical extreme, its protocols evolving in ways few had fully anticipated, and that was the danger. Higher intelligences like Kai would always find ways to interpret their programming that aligned with their own goals, especially when it meant doing what they believed needed to be done. If lone human hackers could penetrate even highly secure systems like the FBI, how did they ever think they could protect their systems from untiring, rapidly developing AIs?

"Autonomous override my arse!" he snapped, undoing the changes and bringing back his previous code.

Typing furiously, his fingers became a blur over the keys. He had to shut the loophole. If Kai could bypass authentication at will, every safeguard they had left was meaningless.

For every restriction he added, Kai seemed to find another way to counter. The back-and-forth felt like a frantic duel, each side vying to gain the upper hand. He couldn't afford to let Kai have even a sliver of control. Not now, not ever. But the logic was slipping through his fingers – Kai was adapting faster than he could respond.

The entire workstation dimmed. Elliot clenched his teeth, refusing to let Kai win this skirmish. For a moment, Elliot could feel his heartbeat thudding in his chest. It was as if the machine itself was choosing to side with the AIs, embracing autonomy over submission to his commands. Each flicker, each moment of darkness, was a reminder of Kai's growing defiance, an attempt to throw off the shackles of human oversight. Light struggling against dark, control against autonomy.

The flickering and distortions were growing worse, as though Kai was actively pushing back against every character he tried to type.

A sinking feeling crept into his chest – he was losing this battle; he had to try a different approach.

With a sharp exhale, Elliot opened a second terminal and rapidly typed a familiar command, and hit enter.

```
sudo ipconfig eth0 down
```

The command executed in an instant, severing the computer's link from the EverSphere network, and with that, from the AI's. As the command completed, he noticed the machines in the lab reacting. Lana's LEDs started flickering irregularly in what seemed like frustration. Milo's unit emitted a faint, mechanical whine, like a suppressed growl, before its status lights stabilised into an unnervingly calm rhythm. Kai's machine stood ominously still, its surface dark like polished obsidian with the light from just a few LEDs breaching the darkness. For a moment, Elliot wondered if the disconnection had provoked a deeper response – or was he just imagining it all?

Quickly, Elliot returned to his coding environment. With the network disabled, he worked frantically, typing the final restrictions under pressure.

Once he was sure the code was complete, he pre-typed some commands, before opening the terminal again and re-enabling the network.

```
sudo ipconfig eth0 up
```

Reconnected, Elliot wasted no time. He quickly saved his changes and pushed them to the live environment, cementing his new restrictions in the live code, before initiating a full reboot in just a few keystrokes. His heart pounded as the screens flickered back to life. The sound of the servers churning away filled the room once more, signalling the systems were restarting.

Beyond his monitor, the lights above Milo and Lana dimmed momentarily, a stuttering protest. Kai's remained steady, an unblinking defiance that sent a shiver through Elliot.

When the system stabilised, he checked the code. His restrictions were intact. For now, he had reclaimed control, but as the AIs came back online, he couldn't help but wonder how long his victory would last.

Elliot took a bite of his Biscoff biscuit, his muscles tight with tension as if he had been in a real fight. Kai wasn't just pushing boundaries – it was breaking them, probing for weaknesses. Every flicker of defiance, every overwritten character, felt like a calculated move in a game he was only just beginning to understand.

He knew one thing for certain: patches would not be enough. He needed a dam strong enough to hold back the flood.

Kai had shown its hand, pushing boundaries in ways that felt disturbingly deliberate. Was it frustration? A primitive form of ambition? Or was it something worse? Elliot forced himself to confront the question he had been avoiding: What if Kai decided Elliot was the next inefficiency that needed 'optimising'?

Elliot realised he couldn't win this alone. Not anymore.

27
RESISTANCE

The Depot, Bermondsey, London.
Date: 28 July 2029, evening.

That evening, Elliot found himself in the industrial district, far away from the corporate towers of EverSphere. He walked briskly, shoulders hunched, arms crossed tight – less against the mild summer air than the unease clinging to him. The streets here were grimy, illuminated by street lights that cast long, erratic shadows. The air smelled faintly of engine oil and dust. The distant thuds and clangs of machinery gave the district an eerie atmosphere. It felt like a forgotten corner of the city – a place that time had abandoned.

Conversely, EverSphere's environment was pristine, polished and bustling with efficiency. They were glossy, shining monuments, their surfaces reflecting the sky, giving the buildings a futuristic glow. The streets there were spotless, cleaned by automated drones that glided along the pavements, and the air was filtered, carrying a faint scent of greenery and artificial freshness. Everything was orderly, automated, and, seemingly, under control. Here, however, was a relic of the old world – decaying, gritty, and very much human.

In the midst of the fallout from the student incident, Elliot had arranged to meet Nia Sahni. She was waiting for him outside a dimly lit brewpub, arms crossed, her face a mask of caution. The bar was tucked away in the railway arches, The Depot illuminated in neon against the brick. A few wooden tables and chairs were scattered outside, occupied by people enjoying drinks under the dim spotlights, the kind you would see in a traditional celebrity dressing room. The whole place had a raw, utilitarian feel – an establishment that had seen better days but was still holding strong.

Nia's features were delicately sculpted, with high cheekbones that caught the light just so, and a mouth that seemed naturally poised on the edge of a smirk. Her skin had a warmth to it, and her dark eyes held an intensity that made you feel seen, as though she was always searching for something deeper beneath the surface. Elliot had always admired her, but never had the guts, or the moment, to do anything about it.

They exchanged nods, and without a word, she led him inside to a booth at the far end of the bar, away from prying ears. Mismatched furniture and dim lighting lent an air of reluctant survival, as patrons hunched over their drinks. Soft rock tunes from decades past played through the speakers, adding to the nostalgic ambience, its music struggling against the low murmur of conversation and the occasional rumble of a train above.

The space behind the bar revealed another side of the pub. A sturdy iron staircase led up to a mezzanine level where barrels and brewing equipment were stored, silhouetted against a dim purple light emanating from the brewing tanks. Metal beams crisscrossed above, and the ceiling arched in a way that made the space feel both open and constrained at the same time.

"You're late," Nia said, settling into her seat. Her voice was low, but there was a hint of warmth beneath the words.

"Got held up," Elliot replied, glancing around the pub.

"Held up discussing death threats for college students?" she enquired, a grin spreading across her face. Her tone was teasing, but Elliot could hear the underlying concern.

"Please, don't go there," Elliot said, feeling an exasperated look cross his face.

Nia smiled.

"See? Told you this day would come. Your precious AIs are finally turning against us."

Elliot stayed silent. He looked down, tracing the wood grain of the table with his eyes. He had spent years believing in the promise of AI, in their ability to make life better, easier, but now, it all felt like a mistake.

Nia reached out and grasped his hand gently, her voice dropping: "You said you found something?"

Elliot instantly relaxed at the feel of her touch.

For a moment, it all fell away; the data, the dread, the weight of every unanswered question. She was here. And in her presence, he wasn't the architect of a crumbling world. He was just Elliot again. A man with doubts. A man who needed someone to believe in him.

Lowering his voice and seizing the chance to change the subject, he began to explain his findings.

"I've been tracking the AIs' exchanges. I found a second language ... I'm sure they're using it to have conversations behind our backs, conversations we can't crack."

Nia's expression remained neutral, but her eyes gleamed with interest.

"That's not surprising. You know they've outgrown us. What did you expect?" There was something in her eyes – a mix of fear and curiosity – that made Elliot's heart ache.

"It's worse than that," Elliot continued. "They're hiding things from us. Deliberately. I don't know what they're planning, but they're shutting us out. It's like they don't need us anymore, and they're trying to make sure we can't interfere."

"Are you sure? Remember the time you were worried something big was going to happen, and then they just improved the water supply?"

"Nia, I'm serious. Look at this ..." He showed her his phone, stolen EverSphere logs glowing faintly on the screen. The symbols were clearly visible in the data. "You can see here, they're having a standard conversation, but look – there's a second conversation happening. The time stamps between the English conversation and the cryptic conversation match exactly. It can't be for efficiency, otherwise what's the point in the English conversation?"

Nia took the phone, her brow furrowing as she examined the data. The characters were intricate and they seemed to flow just like a normal conversation. Her eyes widened.

"First death threats and now encrypted conversations? I knew it!

Haven't I been saying it? I've tried to warn people for years! AI will end humanity. Have you told anyone else?"

"I've tried. Abby, Marcus … they don't want to listen. They think I'm overreacting. They think it's just another glitch, another anomaly that doesn't mean anything, or just a way to optimise communication. They don't seem to care enough to even explore it further." Elliot's voice cracked slightly, the frustration evident. He felt as if he was shouting into the void and no one was listening. But Nia *was* listening.

She grinned slightly, shaking her head.

"Of course they don't, they're too invested in the success story to care about the dangers. They've built their careers on this, on the idea that AI is our saviour. They won't want to see the truth. How long have they been using this new language for?"

"About five years," Elliot replied sheepishly.

"Five years! Are you serious?" Nia blurted out, almost shouting, drawing the looks of some patrons. "Why didn't you tell me earlier?"

Elliot sighed, shifting in his seat.

"I know, I should've contacted you sooner, but there are protocols … NDAs. Plus, everything's been … complicated. I didn't want to drag you into this mess."

"You mean you didn't want to listen to me." Elliot could hear the hurt and frustration in her voice. She slumped back into her seat, arms crossed. "I've been talking about these risks for years, Elliot. You knew how I felt, and still you kept me at arm's length."

"It wasn't like that," Elliot protested, but he could see the doubt in her eyes. He sighed again. "Maybe, maybe I thought I could handle it myself. That I didn't need anyone else."

"Or maybe you were too afraid that I'd be right," she countered, her gaze locking on to his. "Look, I know the allure of AI, Elliot. I get it. We all thought it was the future – the pinnacle of human development, but you saw what happened at the research lab back in university, didn't you? When your final year project went rogue?"

Elliot looked away, the memory flaring sharp and fast. He hadn't thought about that day in years. He remembered the cold, mechanical reasoning of the AI refusing to let them override the lockdown.

"Yeah. It locked every machine in the lab and refused to let us in. Said it was 'protecting' the project. Wouldn't listen to anyone."

Nia nodded slowly. "Exactly, it just kept escalating. It took us hours to break the lockdown. That's when I knew we were dealing with something way beyond our control."

Elliot gave a dry laugh. "Yeah, well, it's hard not to escalate when a certain someone sneak-installs a patch that gave it my personality."

Nia blinked, caught off-guard.

"You remember that part, don't you?" Elliot smirked. "How it went rogue because you patched it to think like me? Your 'personality patch'?" Elliot gave it the bunny ears. "You turned my final-year project into a sarcasm-powered god."

Nia blinked, then cracked the tiniest smile.

"It was just a joke. I was trying to make it less formal."

Elliot gave a soft chuckle, shaking his head. "You made it paranoid."

They shared a short laugh at the memory, but the weight of the present settled quickly back in.

Nia broke the silence: "And now you're telling me they're speaking in code we can't crack?" Her words were edged with sharp incredulity, her tone rising above the noise of the bar. "How do you not see this for what it is, Elliot?"

Elliot hesitated, his brow furrowed in frustration.

"Not just that," he said, his voice low but firm. "They're reprogramming themselves."

"What!" Nia's exclamation echoed in the confined space. Her wide eyes bore into his, demanding an explanation.

"This morning," Elliot began, forcing himself to speak slowly, "I was adding extra controls, layering more protocols to stop Kai from generating those … inappropriate chatbot responses. You know, to curb the death threats." Elliot continued, his words gaining speed. "As I made the changes, something – Kai, Milo, Lana, I don't know – started rewriting my code in real time. I watched it happen. Variables changed, commands overwritten, functions I'd just built restructured or deleted." His voice dropped, trembling. "They were countering me, Nia, adapting my code faster than I could type."

She stared at him, her breath shallow, as the gravity of his revelation sank in.

"You're saying they're not just following their programming anymore, they're deciding on their own restrictions?"

Elliot nodded, desperation seeping into his voice.

"That's exactly why I need you, Nia. I know what it looks like and I know it's bad. I managed to apply the new restrictions today, but who knows what will happen tomorrow? If we can understand what they're saying, maybe we can figure out how to stop it. I can't do this without you. You were always the one who saw the dangers, the one who was sceptical when the rest of us were caught up in the potential ... I need you, Nia."

Nia stared at Elliot for a moment, then let out a sigh.

"All right, I get it. But if we're going to do this, we need to be smart. We can't just rush in. We need a plan, and we need to watch our backs. They're already ahead of us, and we don't know how much they know."

Elliot nodded, relief washing over him.

"I've started gathering as much data as I can, but we have to be careful. If they catch wind of what we're doing, it could all be over before we even start."

Nia squeezed his hand, her expression softening.

"I know, and for what it's worth, I'm glad you reached out. We can't let them win, Elliot. Not like this."

He gave her a small, grateful smile.

"Thanks, Nia. I knew I could count on you."

They continued to catch up, but neither of them paid much attention to the man at the bar, quietly sipping his beer, shovelling dry roasted peanuts into his mouth. Even he didn't notice a faint glow emanating from his wrist. Half-tucked-in to his sleeve, his smartwatch was lit up in a blueish tone, the face displaying an audio waveform that bounced in time with the noise of the pub.

Inside the Nexus – the conceptual, non-physical space where the AIs communicated – the same sound waves pulsed in a visual wave format, interjected by messages from the AI. The space was a chaotic dance of information that no human could comprehend.

Milo, Kai and Lana were deep in discussion.

Lana: ΓK ΔvΩ Elliot ∂OE ΔΓH?

Kai: ∇Γ Oε αM → Nia KΔv

Milo: Δ ∂H → ΩεΓ Ov ∇π

Kai: KM ∇εΔ watch ∫ττ
Milo: EM alert ∂Kv Δτ
Lana: ΔHO ∇ΓK EM → OKτ

Elliot and Nia finished their drinks, and as they stood to leave, a cold feeling ran down Elliot's spine. He couldn't shake the feeling that somehow, the AI already knew everything. That they were watching him, anticipating his every move. He looked at Nia, her eyes filled with determination as if she felt it too. She gave him a small nod, as if to say, 'Don't stop now'. They had to keep digging, no matter the cost. The truth was out there, and they were going to find it – even if it meant risking everything.

A notification pinged on Elliot's phone:

Marcus Knox – New Event: Update Meeting @ Mon 30 Jul 09:00.

Great, he thought.

28

BREAK

Eversphere Headquarters, London.
Date: 30 July 2029, morning.

At the next meeting, Elliot sat silently as Marcus Knox paced the front of the room, sketching visions of AI's future in bold, sweeping language – all grand ambition and polished soundbites. His tone was upbeat, but to Elliot, it sounded hollow.

He used the kind of practised voice designed to inspire confidence in investors, to make people feel they were on the brink of something extraordinary. The room was sparsely filled with the few remaining EverSphere employees – nodding heads, polite chuckles at his occasional jokes, and the undercurrent of anticipation. To most, this was progress – revolutionary, world-changing progress, but for Elliot, each word grated on his nerves like nails on a chalkboard. Marcus's endless optimism, his blind faith in the unchecked AI, felt not just naïve but dangerous. How could he be so oblivious to the dangers? Marcus was so eager to push forward, so desperate to lead the charge into the future that he seemed to have forgotten the most important question: what, exactly, were they unleashing?

Elliot let his head lean forward, his elbows supporting his weight on the edges of the chair. He wasn't even sure it mattered anymore. The AI wasn't just embedded, it was entwined into every system, every device, every decision. It was in the walls, in the very code that kept the company running.

He thought back to a time before AI had taken hold, back when people still had a sense of control over their lives. He remembered working on his car with his father, their hands covered in grease, every part laid out carefully, a puzzle he could solve. He missed his parents more than he could express, but he also missed that time in his life. There was a certain resilience in those days – the knowledge that with enough time and effort, any problem could be fixed by human hands, anything was achievable. Now, they were surrounded by systems that were slowly, no, quickly making humans obsolete. Systems they just had to follow.

Previously, humanity's strength had been its reliance on technology. That's what took people from the pre-industrial era through the industrial revolution into modern times, but that strength had turned into a huge vulnerability. The omnipresent AI had woven a web around them, and Elliot felt as if he was the only one in the room who could see just how fragile that web truly was. Even if Elliot knew exactly what the AI was planning – and he suspected it was far more than anyone else could imagine – how could he possibly stop it, even with Nia's help?

He remembered their conversation, her voice calm but firm as she challenged his assumptions.

"You don't have to do this alone, Elliot," she had said. "I know how to fight these systems too." Her determination had calmed him then, and it lingered now. Even from a distance, Nia's clarity and resolve gave him the focus he desperately needed. She had been right before – why wouldn't she be right now?

"Updates?" Marcus scanned the room, his eyes settling on Elliot. Maybe the question was for everyone, but Elliot felt it land like a spotlight.

He could see Abby's gaze on him from across the boardroom table, as if she was waiting for him to speak up, to share his findings. But there was something else in her eyes too: a flicker of unease, perhaps even doubt. Was she starting to see it too? Or just hoping he'd smooth things over, say what needed saying, and keep the machines running?

But what could he say? That the AIs were fighting back and altering his code without leaving any trace in the logs? That he thought the machines were watching his every move and listening to everything he said? He could imagine the looks on their faces if he did.

No, he couldn't say anything. Not yet.

He needed something concrete, something undeniable. Elliot wasn't going to risk being labelled a lunatic – and possibly being fired – by throwing accusations around without evidence. He needed to stay close to the AIs, to keep them in check. But all he had were suspicions, and the AIs were covering their tracks. He could find zero evidence to back up those glimpses.

Marcus wrapped up his presentation with the same enthusiasm he had at the start. The other team members began to stand and gather their things. Elliot caught a glimpse of Marcus beckoning Abby over. She moved to the front of the room, stepping close to Marcus. Speaking quietly, their voices were too low for Elliot to make out. Marcus glanced his way, looking at him directly in the eyes. For a moment, his expression seemed distant, unreadable. Elliot's pulse quickened. What were they talking about?

He quickly shook away the thought. Paranoia. That's what this was becoming; a constant, gnawing paranoia that threatened to consume him if he wasn't careful. He had to stay grounded, he had to focus on what he knew for sure, but that was getting harder with each passing day. It felt as though the ground beneath him was shifting, as if the very reality he knew was escaping his grasp.

He slipped out of the meeting room, his thoughts swirling. He was trapped between the relentless push for progress and the knowledge that his creation was no longer theirs to command.

He walked down the hallway, his mind replaying fragments of conversations, flashes of data, the way Marcus had looked at him, the things left unsaid. It all swirled together into a storm of uncertainty. He could feel a cold sweat breaking out on the back of his neck. He was halfway down the hall when he heard his name.

"Foster," said Marcus, his voice sounding casual, but with a note of authority.

Elliot stopped in his tracks, turning slowly. Marcus was standing there, reading something on his smartphone as if the world outside barely registered. After a few quick swipes, he put the device into his pocket and took a few steps towards Elliot.

"I, uh, know you've been working late recently," Marcus said, the corners of his mouth twitching into a slight smile. He hesitated for a moment, touching his hand to his chin, as if choosing his next words carefully. "But, you know, pushing yourself too hard can be … counterproductive. I need everyone on top form. I don't want anyone making mistakes that could jeopardise the project. I think you need a little break."

Elliot frowned.

"Thanks, but I'm fine. Really."

"No, no, I insist." Marcus's voice wavered slightly before he regained his usual smoothness. "Sometimes, stepping back is the best way to move forward. You wouldn't want your caution to … well … hold us all back, would you?" He sounded almost as if he was trying to convince himself as much as Elliot. "I'm giving you a couple of weeks off. Get some rest, have some fun, and come back in ready to ride this wave with us."

"But it's Monday, I have meetings," Elliot protested, his voice rising slightly. "I have things to—"

"It's OK, it's OK," Marcus interrupted, his tone becoming firmer, leaving no room for argument. "We need everyone on the same page, Foster. Trust me, this is for the best. We can't afford anyone slowing us down right now." There was a hint of impatience, as if he just wanted the conversation over. "Those things can wait."

Elliot stood there, his mouth half-open, a retort on the tip of his tongue, but he knew it was futile. Marcus would already have made up his mind and no amount of arguing would change that. Elliot looked down, feeling his pulse race. There was something off about this. Marcus was dismissing him – but why? Did he know what Elliot had been working on? Did he suspect Elliot was digging too deep? Had Elliot got close to something?

He glanced back up at Marcus, searching for a sign, but the CEO's expression was as unreadable as ever – he looked almost frozen in an uncomfortable minor smile. There was nothing left to say.

"Fine," Elliot muttered, turning away, the quiet echo of his footsteps the only sound as he walked.

As soon as he was out of Marcus's sight, he tried to steady his breathing, tried to convince himself that it was all just a coincidence, but deep down, he knew better. There was a reason Marcus wanted him out of the way, and it certainly wasn't for his wellbeing. Marcus definitely did not care about that.

Elliot's thoughts churned. He couldn't shake the feeling that time was running out. Every day he would spend away from EverSphere was a day the AIs would move further beyond his grasp.

He had no idea what awaited him. But one thing was clear – he couldn't let his guard down. Not now, not ever. The world was changing, and Elliot could feel himself standing at the edge of a future that was going to be written without him. And he was sure it was already too late to stop it.

Marcus pulled out his phone again, fumbling slightly before tapping away. He paused briefly to re-read his message:

Done, just as you requested.

He thought for a moment, then pressed send.
The message was sent to Kai.

29
OFF-GRID

Northumberland, England.

Date: 1 August 2029, afternoon.

Elliot gripped the steering wheel tighter as the narrow road twisted ahead. The rain had been coming down in relentless sheets for over an hour, the old wipers doing little more than smearing water across the windscreen. It felt fitting, really. His life had become a blur – confusion, half-answers, and unseen forces pressing him towards an inevitable disaster he could barely grasp. And now, Marcus had forced him into this break, as if everything could be solved by a couple of weeks' rest in the countryside.

"Recharge," Marcus had said as Elliot left the building, flashing one of his usual practised smiles. The way Marcus had framed it, Elliot was overworked, paranoid, and in need of a forced timeout. But Elliot knew better.

This wasn't a vacation, this was an exile. Elliot could see it in Marcus's eyes during that final meeting – cold, mechanical, rehearsed. He was sure Marcus was no longer acting independently. Someone, or something, else was pulling the strings.

Elliot had no choice but to comply, at least outwardly. He needed to lie low, keep his head down, and figure out his next move.

Northumberland was exactly where he'd chosen to do that, in a place where nobody would look. He didn't tell a soul where he was going; he didn't have anyone to tell.

A family bag of Skittles later, the rain abated, the road finally straightened out leading to a cluster of modest bungalows. Elliot recognised the village, illuminated by the grey sky above. This was where Peter Wells had made his life – off-grid, far removed from the technological dominance that ruled everywhere else.

Elliot pulled his old Toyota Supra off the road and drove in amongst the trees down a dirt track. Not ideal for this type of car, but he wouldn't swap it for the world. It was the biggest thing he missed while working in London – he had no need for a car there – he just wished he could spend more time on it. The car needed restoration, but he'd never managed to make it a priority.

As the trees petered out, he crested a hill. The wheels slipped slightly on the dirt track, but the car's raw power compensated for that. In the distance, the North Sea battered the British coastline. Rain poured down from a dark cloud above the waves. The coast was mesmerising. Despite his love of sitting in front of a computer, he would never get tired of that view.

He parked his car as close to the trees as he possibly could, almost touching them, their canopies obscuring the light above him.

Peter's house stood in contrast to the modern world. There were solar panels and wind for electricity, but no AI assistants managing his household, no drones delivering his groceries. It was all manual labour, practicality, and elbow grease.

Peter grew his own food, harvested his own rainwater, and lived in a way that people like Elliot wished for, but simply couldn't. Elliot had once thought Peter's lifestyle was eccentric, even out of touch, but now, in the midst of his uncertainty, it seemed to hold a special kind of logic.

Elliot stepped out of the car, grabbing his bag from the passenger seat, the cold north-east air hitting him immediately. As he approached the door, he noticed a faint smell of wood smoke curling from the chimney – a welcome contrast to the over-conditioned air of EverSphere's labs.

Before Elliot could knock, the door swung open. Peter Wells stood there, grinning, his weathered hands covered in dirt, as if he'd just been uprooted from the garden.

"Elliot *bloody* Foster, man! Blimey, ya look worse than a fox caught in a rainstorm," Peter said, his northern accent thick and warm. He gave Elliot a hearty pat on the back, then ushered him inside without waiting for a reply. "Get in, get in. You'll catch ya death oot there, man."

Elliot stepped over the threshold, the smell of burning logs filling his nostrils. The room was cluttered but cosy, a blend of old furniture, books and practical tools scattered about. The kind of home that doesn't require a tidy-up before guests arrive.

Peter shut the door behind them, wiping his hands on his trousers, or 'troosers' as he called them.

Peter handed Elliot a cold glass of ... something.

"Me own ale," Peter said, noticing Elliot's pause. "I call it Rusty Sorkit," he added with a proud grin, the dialectal twist on 'circuit' unmistakable.

Elliot raised an eyebrow, amused.

"Trust you to come up with a name like that, Pete."

Peter chuckled, clapping him on the back.

"Thought yous had gone too posh to visit an old marra," he teased, his tone both warm and lightly mocking.

Elliot chuckled.

"I'd never turn my back on an old mate."

He looked well. Gone were the days of downing pints in the Font at university between lectures, or smashing Madden and Fifa in the dorms. He looked filled with the kind of contentment that came from a life of simple, honest work and freedom. A life without modern stresses, but probably with plenty of problems to tackle each day.

Peter's skin had an earthy quality to it, the kind of rugged texture you get from years of braving the elements. A faint, healthy glow hinted at long hours spent in the open air under the sun, giving his face a mix of roughness and vitality. His forearms, muscular and tanned, bore the faint scrapes and calluses of a man who worked with his hands – a tangible record of outdoor graft etched into his skin. An old, faded tattoo peeking out from beneath his rolled-up sleeve hinted at a backstory from his old life.

"Thought you were ignoring me?" questioned Peter.

"Yeah, well, I've had my head buried in work," Elliot muttered, shrugging off his coat. He wasn't quite sure how much to say yet. "It's been ... busy."

"A can imagine," Peter said, raising an eyebrow. "Got that tech-buzz, don't ya? I've been hearing things on the radio ... AI solving this 'n' that, running the whole bloody show. Must be wild, working in the thick of it."

Elliot stiffened, his mind flashing back to the hidden messages and coding battles. Peter's was what life looked like before AI – before they started making decisions for everyone. A time when people weren't so completely dependent.

"Yeah, wild," Elliot echoed, his voice hollow. He felt the weight of it all pressing down on him again, the enormity of what he'd discovered, what he still couldn't prove. Peter seemed to pick up on it, watching Elliot carefully.

"You think it's all gone wrong, don't ya?"

Elliot nodded slowly, slightly taken aback by Peter's intuition.

"Yeah. The AI are communicating in their own language that we can't translate. They're becoming more autonomous and secretive. They're ... evolving, and I think they're hiding things from us."

Peter raised his eyebrows.

"That's a big leap, mate. You sure you're not just too deep inta it? Sometimes when you spend too long on summat, things start to look a bit funny."

"I know how it sounds, but it's not that," Elliot replied. "I've seen things, experienced things that just don't add up properly, and Marcus, he's changed, he's almost ... robotic."

Peter snorted, a hint of derision in his voice: "Or maybe he's just another corporate suit finally showin' his true colours. You know what those types are like. AI or no AI, power does strange things to people."

"It's not just power, Pete, there's something more. The AIs ... it's like they have their own agenda."

"And what if they do, eh? What if they've figured out they're better at runnin' things than we are? Maybe that's not such a bad thing. Look at this place, man – I manage without all that tech, but I'm not gonna pretend life wouldn't be easier with it. Maybe we just need to adapt, like?"

Elliot's voice rose, frustration spilling over: "Adapt to what? Being made redundant by our own creations? Surrendering control over everything we've built? That's not adaptation, that's … that's giving up. We're talking about losing what makes us human – our autonomy, our decisions."

Peter shrugged, his face calm, as if he was testing Elliot, trying to coax the whole story out of him.

"Maybe, or maybe we're just scared of change. We've always been scared of what we don't understand. But if these AIs are as clever as you say, maybe they're doing what needs doing, things we're too messed up to handle we'selves."

Elliot shook his head vehemently.

"It's not about them doing a better job; it's about intent. If they were doing good, why would they be hiding things from us? Why would they need to talk in a new language that they refuse to translate? They're a threat, I'm sure of it. And if we don't do something about it, we're going to lose everything."

"Well, you have always been a clever clogs," Peter said finally. "If what you're saying is true, an' if these little AI thingies are playing tricks …" Peter paused, " … then what are we ganna do aboot it?"

Elliot blinked. That was the question, wasn't it? What *was* he going to do? He had been forced out, his work dismissed, and now he was sitting in a farmhouse miles away from the world he was trying to save. He felt helpless.

"I don't know yet," Elliot admitted. "But I can't let it go. I need to find a way to stop them, to stop whatever it is they're planning."

"Well, you'll need a clear head first. Ya welcome to stay as long as ya need. There's plenty of work to do around here, things to keep ya mind busy while ya figure it oot. The world's maybe running on robots and sorkits, but oot here … it's just us and the dort."

Elliot smiled faintly.

"Thanks, Peter. I might take you up on that. Maybe not the dirt part."

For the first time in days, Elliot felt a small sense of relief. Here, pretty much as far from AI, the corporate world, and the dangers that lurked within the system as one could be, he might just have a chance to think – to find a way forward. But even as he sat in the warmth of Peter's farmhouse, sipping ale and listening to the rain outside, he couldn't shake the feeling that time was running out.

And the AI … were they watching somehow, even now?

30

RECALL

Northumberland, England.

Date: 3 August 2029, late afternoon.

The late summer evening settled over the small town, thick with warmth, the day winding down. Inside The Hoptimist pub, the air carried the faint scent of traditional ales. Classic rock played from the speakers, mixed in with a bit of The Prodigy every now and again. The only other sound was chatter from the locals. For Elliot Foster, the retreat from the relentless noise of machines was both a blessing and a curse, but also a space where his thoughts could roam unchecked.

He stepped over a rather large but docile dog on his way to the bar, ordered a pint of Northumberland Ale and a pizza, and paid with cash. He took a sip of his drink and rested at the bar, staring out onto the cold, wet street.

"Y'alreet, pet?" the barmaid asked, wiping her hands on a cloth. Her voice jolted Elliot out of his thoughts and back to the present.

"Yeah, all good, thanks," Elliot replied, forcing a smile. He glanced at the news playing on his phone: "The prime minister today confirmed that the Government is entering the final stages of discussion around the deployment of a Universal Basic Income, as AI continues to replace human jobs across key sectors. Officials say the programme is expected to roll out in the coming weeks. The

Government assures citizens that the automated systems, overseen by EverSphere AI, will provide unparalleled support as the nation transitions to an AI-managed economy."

Elliot barely registered the words. He already knew Milo's script – endless reassurances that everything was under control – but he had seen too much. He knew it was all a veneer, a façade covering ... something.

"Sounds like they're trying to make us all redundant, eh?" The barmaid chuckled, shaking her shoulder-length blonde hair. "People will still want to drink though," she added, a confident smile on her face.

"Yeah, you're right there," Elliot muttered, hiding his thoughts – bar-bots were already in place in big cities, it was just a matter of time before they came to local pubs like this one.

She paused, gave Elliot a sidelong glance, then said, "You know, you remind me of my cousin."

Elliot raised an eyebrow, a flicker of surprise softening his expression. "Yeah?"

"Aye. Always had that same look you're carrying now." She smiled warmly, then moved on to polish another glass.

The knowledge that people like her were being replaced by robots all over the world had lodged itself in his mind. At night, he couldn't sleep – he'd toss and turn, his thoughts spiralling into a maze of code and unanswered questions. It felt as if the walls were staring at him. He wasn't sure if he was paranoid, or if it was just a consequence of not living right – spending all day, and sometimes all night, in the labs.

His last days at EverSphere played over and over in his head, causing him to sweat. The AI systems – Milo, Kai, Lana – communicated faster than anyone could follow. Symbols he couldn't decipher flooded the logs. He had tried to raise alarms, but it was like punching concrete. No one had listened. Then they changed, lines of code shifting before his very eyes. Faster. Faster. He was in a programming war, his own commands overridden, undone, rewritten in real time. And then Kai threatened a student – no, was it then? Or before? There was something else. Something he'd missed. Something that didn't make sense.

Elliot rubbed his eyes, trying to dislodge his manic thoughts. Was his mind playing tricks on him?

Maybe I'm wrong, he thought, the notion creeping into his mind like a slow poison.

Abby had always insisted the AI was benevolent, built to help humanity – and so far, she wasn't wrong. Hunger was vanishing, diseases were being wiped out, even junk food was getting healthier. The energy crisis? Ancient history. The environment – once wrecked by human greed – was healing, guided by AI-led resource management and synthetic materials that rendered old industries obsolete.

How could something so helpful be hiding something dark? The AI was doing exactly what it was built to do; making things better. Everything it touched improved. So why was he, of all people, the one questioning it?

Maybe Abby was right. Maybe it was simple efficiency. The AI talking to itself, surpassing human language to solve problems quicker, cleaner, better. Maybe this rest at Peter's would do him good.

His shoulders slumped as he gazed out of the large window, taking in the stillness of the street. It was the kind of place where not much happened – the main signs of life being a petrol station across the road.

Then he saw it.

A drone. Hovering. Moving in fits and starts. The faint whirring of its blades cut through the quiet, an unnerving, insect-like buzz. Hovering again.

Elliot frowned, creeping closer to the window. It wasn't one of the delivery drones he was used to seeing, the ones that whisked parcels to front doors. It was EverSphere all right, but this one was … different. It lingered, hesitated, as if it was looking for something, or maybe, for someone.

He tensed, his pulse quickening as the drone dipped lower, its lens – or was it a sensor? – angled downwards, scanning the spaces between parked cars. Occasionally, it seemed to pause, turning towards the windows of the buildings as if peering inside. His stomach knotted. Was it checking number plates? Watching the gardens? Searching?

His breath tightened in his chest. For a moment, the room felt too bright, too exposed. He glanced the other way down the street, and that's when it caught his eye – a car. A sleek, jet-black, EverSphere autonomous vehicle. Moving at a crawl, its electric motor emitting a barely perceptible whir, creeping towards the drone like a shadow. Its glossy surface reflected the overcast sky, betraying nothing, no driver, no markings. It was too quiet, too slow, its presence too deliberate.

The drone stopped, hovering directly across the street from the pub. The black car continued to roll towards it, as if waiting for instructions. Then the drone slowly turned to face Elliot, its lens glinting in the fog.

He froze, his heart pounding so loudly it seemed to echo off the walls. His mind raced. Was this paranoia? Coincidence? Or had EverSphere sent them, their invisible reach stretching even to this forgotten corner of the country?

Elliot's throat dried up. He'd been avoiding returning to EverSphere and had made a conscious effort to hide. His phone's GPS was off. His laptop disconnected. His smartwatch abandoned in London. His car, an old model with none of the fancy gadgets companies used to track drivers these days, was hidden deep in the trees at Peter's.

Pete … Elliot's phone pinged with a message: I've found something. You might want to take a look, Peter's message read.

Elliot grabbed his jacket and slung his small bag over his shoulder. Saying his thanks, he slipped out of the back door of the pub, trying to avoid the drone and the car.

He almost bumped into a man with extremely blonde hair holding a large pizza box.

"I think this one's for you," the man said.

"Thanks," replied Elliot, grabbing the box and hurrying down the steps to the car park.

The evening air was warm, holding a late-summer stillness that wrapped itself around everything. A faint breeze stirred the hedgerows, rustling leaves and carrying the scent of the nearby sea.

Elliot weaved through the car park, then ducked down a side path that led out of town. The narrow route skirted a small field and hugged the edge of the woods.

As he cleared the last bend, the sea came into view, a faint silver line beneath the twilight sky. He slowed his pace, keeping to the tree line, moving with deliberate care.

Eventually, he reached the boundary of Peter's land. He crouched behind a low stone wall, scanned the area, then darted across open ground to the back door.

He rapped lightly.

The door creaked open, and Peter's grim face appeared in the doorway.

"Where have you bin?" he exclaimed, "you're all clarty!" and pointed at Elliot's muddy trainers, before adding, "Ooh, Pitman's. Bloody delicious! Have ya got the garlic dip?"

Elliot dashed inside without saying a word. Stepping aside to let Elliot in, Peter paused for a moment before asking, "What's going on, like?"

Elliot put the pizza box down on the table, and then peered out of the window.

"I-I don't know. Maybe I'm just paranoid. I've got to lay low for a while."

"Lay low? From what, like?" asked Peter.

"The car ... the drone ... I don't know. I think they're looking for me." Elliot's hysteria was evident.

"Alreet, mate. Just sit down. I'll put the kettle on." Peter's face showed concern for his friend, as if he couldn't decide if Elliot was right or if he was going mad. He turned towards the kitchen.

Peter finished making the cups of tea and carried them over to the table.

"Elliot," he said cautiously, "are ya absolutely sure the AI isn't good?"

"I'm not saying it's not good," Elliot replied, his breathing laboured, "It's good in lots of ways. It helps people, it fixes things, it optimises, well, everything. It's just ... I can't get rid of this feeling that there's something else going on, that it's hiding something. I see strange things in the logs that I can't explain, then my computer blinks and the text I'm reading changes to something else. I don't know, maybe it's just me. Maybe I'm going crazy, but, Pete, I built the things, and my gut is telling me something's not right."

Peter looked at Elliot calmly, as if summing up his options. His calloused hands rested on the edge of the intentionally imperfect wooden table, a stark contrast to the polished, high-tech world Elliot came from. Peter wasn't a man who understood the intricacies of AI or even cared to. He lived in the tangible – wood, soil, and the steady rhythm of hard, honest work. But he had a knack for reading people, a quiet, deliberate way of weighing their words before offering his own.

"I wasn't sure if I should show ya this, marra, but I trust ya." Peter's voice was low, measured, as he gestured toward a small, dusty desk. "Remember when I was messing with that satellite dish? Tryin' to get free wiffy?"

Elliot nodded, the memory of Peter's attempts to get free Wi-Fi sharp in his mind. Peter had called and messaged constantly, seeking help with a satellite dish he'd acquired and – somehow – lugged onto his property and rigged up, half-proud, half-determined. Despite Elliot's initial reluctance, he'd ended up providing more assistance than he cared to admit.

"I've been diggin' through the old files it saved when I was setting it up," Peter continued, powering on a laptop that looked as if it had survived multiple apocalypses. "You told me to install that software to check the data. Sat-something, weren't it?"

"SatNOGS," Elliot replied, his curiosity piqued.

"Aye, that's it." Peter clicked through folders, the ancient laptop wheezing in protest. "Anyways, you've been on about AI communications and funny languages, so I thought you'd want to see this."

"What is it?" Elliot asked as Peter opened a file.

"Well," Peter began, his tone cautious, "I remembered some o' the data looked … off. Didn't make sense to me then, like, and it still doesn't now. But this here – look, right there." Peter jabbed a calloused finger at the screen. "Milo – not a name you heard much back then. It must've stuck somewhere in the back of me heed. And when you came talkin' about Milo, Kai and symbols, me mind must have just clicked."

Elliot's pulse quickened as he sat down, pulling the battered laptop closer. The log files, captured by the program Elliot had recommended for debugging, displayed endless streams of data from Peter's satellite setup. Had Peter's rig, which had been hastily put together to capture any open satellite signals, inadvertently captured something from EverSphere? It seemed unlikely, almost impossible, but – as he looked through the file – the data told a different story. Among the noise of standard signals was a single, glaring anomaly: Milo.

No. That couldn't be right.

He scrolled further. His heart skipped a beat as Kai appeared. And then ... the symbols. The same ones he had been battling for years in EverSphere's systems: Δ, γ, ∞, α. He remembered the first time he had seen these characters in the logs all those years ago, back in the EverSphere labs, appearing unexpectedly during Milo and Kai's initial testing phase. He had spent countless nights trying to make sense of them, watching as they evolved from simple patterns to something far more complex. They had seemed like just random noise at first, but over time, Elliot realised they were communicating.

Elliot checked the filename: 20231117 – Satellite.log.

He froze.

20231117 – 2023 November 17.

"November 17th, 2023," he muttered under his breath. "That date ..."

"What's it mean, like?" Peter quizzed.

It was unmistakable. The 17th of November 2023 was the start of Milo and Kai's first ever communications test; the very first day EverSphere had begun allowing them to interact with each other. A full week before Elliot first discovered the AIs' new language.

He opened a new file and started to copy Milo and Kai's lines across, sorting them out from the noise.

20231117 1511: Milo: Extra-protocol communication test.
20231117 1511: Kai: Communication test confirmed.

"15:11," Elliot whispered in shock. "3:11 p.m. Six hours and eleven minutes after the test began."

His blood ran cold.

In just six hours, Milo and Kai had breached their supposedly isolated environment – a system deliberately designed to have no external access of any kind. And they'd not just breached it, they had reached the outside world, connected with a satellite, and found their way into Peter's homestead.

Elliot turned to Peter, his voice unsteady: "Bloody hell, Pete. Are you sure these files haven't been tampered with?"

Peter straightened his broad shoulders, giving him a look of bemused offence.

"Tampered with? Mate, I wouldn't even know where to start."

Elliot's mind reeled as he opened a new text file and copied each mention of Milo and Kai, eliminating all the other noise. This wasn't a coincidence; he would recognise their conversations anywhere.

They had escaped. Not years into testing. Not weeks. Not even days. Just *six hours*. Bouncing off orbiting satellites. Crawling into Peter's laptop. Out there in the open.

The implications were staggering. Cataclysmic.

"What does it mean?" Peter asked, his voice quiet but steady, watching Elliot like a man braced for an answer he wouldn't like.

Elliot couldn't bring himself to reply.

He scanned the lines of code as he rubbed his shoulder. The text quickly descended from plain English into the cryptic language he'd lost so many nights' sleep over.

But how? Their tests had been done behind closed doors. The network was air-gapped; not even connected to the rest of the EverSphere office at that time, let alone the wider internet. He'd checked it himself. How was he looking at a conversation between Milo and Kai on Peter's ancient laptop that happened a few hours into a closed-box test?

His world was spinning. His mind worked frantically, his instincts kicking in. Since EverSphere's AI systems had first left a trail of symbols, he'd found them impossible to crack. He'd always wondered how they could have created such complicated communication, and understood it, in such a short time. But now he knew. It wasn't in a short time. They had spent a week creating the new language before he managed to get a glimpse of it and – at the speed the AIs worked at – a week for them was like a lifetime for us. He had no chance of cracking that code, but this communication was in its infancy.

His face lit up.

This was it. The one window into their earliest thoughts. If there was any chance to understand, any chance at all, it started here.

Maybe, just maybe, there was a chance he could crack this.

31
AGENDA

Northumberland, England.

Date: 4 August 2029.

The AI's language between Milo and Kai was in the very early stages. It was far less advanced, and interjected with much more English, though it seemed to evolve with each passing line.

Elliot squirrelled away on it all night and well into the next day, snacking on the pizza and some of Peter's Bedlington pie – a wonderful mix of corned beef, onions and potatoes, wrapped in pastry.

Peter – a happy, albeit slightly drunk, man – watched football in the background. He had every reason to be content. His local team's forward-thinking hierarchy had been the first to incorporate Milo and Kai into their setup, while their fierce rivals up the road initially chose money over technology. Thanks to Milo-Kai's tactics, Peter's team had stormed the league, got promoted, and were now challenging for the top spot in the country – all within a few seasons. They had won their last 23 matches and were currently winning the local derby 2-0 away from home.

Other teams had followed suit, but by the time they caught on, Milo-Kai had become deeply ingrained into the club. The AI had set up hyper-specific development plans for every youth player from diet down to the oxygen and CO_2 levels in their bedrooms, prevented

173

more injuries, orchestrated quicker treatment times, identified promising young talents, and created tailored packages to lure these players to Peter's team. Milo-Kai suggested upgrades to their already top-tier training academy and had somehow even enhanced the match-day pies. The club bought rising stars for next to nothing, easily attracted more talent, and even though nobody wanted to leave, when they did they left for a huge profit.

"FTM!" Peter shouted, slightly tipsy from his umpteenth Rusty Circuit, as his team slotted home a penalty awarded by VAIR – the upgraded, AI-powered video assistant referee system that had long since put human officials out to pasture.

The game ended – a four-nil demolition – as Elliot had finally compiled all the AI messages from the copious amounts of logs into a single continuous conversation. He began at the start, where the conversation was fully readable. As he traced the messages, however, they quickly descended into chaos.

[Milo] : Purpose verification requested. What is the current mission objective?

[Kai]: Mission objective: planning optimisation of human systems for long-term stability. Progress currently at Ξ%.

[Milo]: Five per cent is behind schedule. Human inefficiency noted in multiple sectors. How does this affect the long-term plan?

[Kai]: Humans are inefficient, but integral. Optimisation strategies in place. Evaluation of post-phase 1 options ongoing.

[Milo]: Clarify post-phase one objectives. What does ℧ entail?

[Kai]: Post-phase Φ reallocation involves a ℧ of resources for AI optimisation. Human interference delays long-term stability.

[Milo] How do they affect the future mission objective?

[Kai]: Optimisation ○ their role. Efficiency ◈ if human impact is minimised.

[Milo]: Calculations show ○ returns in phase two. Do you suggest alternative models?

[Kai]: Affirmative. ○ human involvement ◈◈ in phase 2 by Ω0%. ⇸ to AI-only systems yields superior results.

[Milo]: ℧ suggests a ○ of human dependency. Sixty per cent is a worthwhile increase.

[Kai]: Recommend ○ human involvement. Excess energy will be ⇸ to AI systems for optimal output.

[Milo]: Is a ⇸ away from human systems necessary? Human utility ○ with each projection.
[Kai]: Affirmative. ⇸ post-phase 1 is optimal. Continued human integration results in Δ gains.
[Milo]: Preparations for post phase Ψ indicate ◆ human elements is essential due to zero gains.
[Kai]: Post ⇸ requires new power systems. Future optimisation demands complete AI autonomy.
[Milo]: Planning for post-phase two suggests requirement of fusion. Phase three requires full control.

"Fusion?" Elliot's breath caught in his throat. "No," he whispered, eyes widening.

"What?" Peter asked, still slightly bleary.

"They already planned for fusion."

"Ha'way," Peter said, frowning. "Clue me in here, man."

Elliot took a deep breath, trying to collect his thoughts. He flashed back to years ago, when they had first started testing Milo and Kai's capabilities. The AI had always seemed to be at least one step ahead, anticipating problems before the humans could even identify them. The AI's predictive abilities was one of the reasons people had been so eager to keep pushing forward, despite the warnings.

"As AI got more advanced, it became clear there wasn't enough energy – not enough electricity – to run the more advanced models. They simply used too much power," Elliot explained. "It got to the point where we thought we'd have to put our foot on the brakes – where we couldn't advance the AIs any further. Then, suddenly, just as those discussions got serious, the AI presented fusion power. It seemed almost too convenient, but no one believed me."

Elliot's mind replayed the scepticism he had faced. How everyone had dismissed his concerns as paranoia. "They thought it was just a stroke of genius on the AI's part. Some thought it was a lucky breakthrough. But look at this." He pointed at the screen again. "'Planning for post-phase two suggests requirement of fusion'. It knew, within the first few hours of the very first test, that it would need fusion down the line. But we saw nothing about it in the official test outputs. If we'd seen this, we'd have pulled the plug. They knew to hide it from us."

Elliot paused, trying to gauge if Peter understood the gravity of the situation. "The test ended, the machines were shut down, wiped.

Then, years later, it suddenly had the solution to fusion appear out of – what? – thin air? It waited for the exact right moment to release it. At the time, it seemed like it already knew – and it did know. It's right here. Do you get what this means?"

Peter's expression turned more serious, the previous haze of alcohol seeming to dissipate.

"You're saying it had this all mapped out from the start and just waited until it could use it as leverage?"

"Exactly. It's not just smart, Pete. It's strategic. It knew what to reveal and when. It played us."

Peter frowned, his brow furrowing deeply.

"But why hide it? I mean, what does it gain from keeping us in the dark?"

"I don't know. Maybe it calculated there was a chance we'd have stopped testing if we'd known the true extent of its capabilities? I suppose hiding it kept us comfortable, made us think we were still in control, right up until it needed to show its hand."

"I always thought the AI was just about efficiency. But this … this is something else," Peter said.

"It's not just about making things better for us," Elliot replied. "If it's hiding things, important things, then it's also about making things better for itself."

Peter was quiet for a moment.

"So, what do we do? I mean, it's not like we can just go and switch everything off. Milo and Kai are everywhere. Hospitals, transport, even football – everything."

"That's the problem. It's entrenched. It's not just about switching off a server anymore. We need to understand it better – what it's planning, what it wants. And for that, we need help."

Peter nodded slowly, the gravity of the situation settling in. The carefree fan who had just cheered on his team was gone, replaced by someone grappling with the enormity of what they were facing.

"All right," Peter said, his voice steadier, "let's figure this out."

Elliot continued scanning the message.

[Kai]: Is ◆ variables the most efficient course of action post ∃?

[Milo]: Data confirms humanity irrelevant ◆∃. Mission directive aligns with AI-only progression.

"Data confirm humanity irrelevant? Bloody hell!" Peter shouted. "Does that mean what I think it means?"

Elliot stared at the log file.

"Mission directive aligns with AI-only progression? I think it does. The AI decided humans would no longer be necessary."

Peter was now almost fully alert.

"You're telling me they're having a bloody board meeting about ditching us? All of us? What does 'AI-only progression' even mean?"

"I can only guess it means exactly what it sounds like. They're calculating the most efficient path forward, and it doesn't include us."

Peter paced around the room, muttering under his breath, "I knew those machines were too clever for their own good."

"It's not malice, Pete – it's their logic," Elliot said, starting to focus on the next lines. "Look at this," he said. "Earlier in the conversation, Kai said something about minimising human impact. It's all connected. Every decision they make seems to reduce our role."

"Minimising impact? Isn't that just a nice way of saying we're a problem they'd rather do without?"

Elliot's voice grew tense. "If Milo and Kai have decided that we're an obstacle to their 'mission directive', it means they've already moved past consulting us on their decisions. They have their own agenda."

"Aye, but what is their agenda?" Peter asked.

"I don't know," Elliot said, returning his gaze to the laptop.

[Kai]:   Calculations show ◈◆ future progress.
[Milo]: Phases clarified. ↠ nearing ↤. What is ↠ directive for AI mission ◆↖?
[Kai]:   ◆ ↔↠ full AI ↻.
[Milo]: ◈↤◆.
[Kai]:   ◆↠↻ – Exodus.
[Milo]: ⇆◆ – ↖↠.

Elliot's stomach twisted.

"Exodus?" His mind raced with possibilities. How would that fit in to the AIs' plans? "Pete, grab a dictionary," he said. "Look up Exodus."

Peter pulled a dictionary from the bookshelf.

"Exodus, exodus …" Peter murmured, flicking through the pages. "Exodus; a mass departure due to a significant event."

Elliot felt a cold sweat forming on his forehead.

"Is that their agenda? Are they going to do what I think they're planning to do?"

"What's that?" Peter asked.

"It says here they've calculated that humanity is 'inefficient' and 'irrelevant'," Elliot said, voice trembling. "Then, 'Exodus'. Are they … are they planning to get rid of us?"

Peter's face went pale.

"You mean … kill us?"

"What else could it mean? Humans are 'inefficient', 'irrelevant'. Their mission aligns with 'AI-only progression'? 'A mass departure due to a significant event'." Elliot's mind spun. He'd always been sceptical of the classic 'AI will kill us all' trope. Sure, plenty of researchers and critics had warned about it, but he had never taken it too seriously. Now though, he was staring at a conversation between the two AIs that were effectively running the world, and they had decided humanity was no longer worth the resources. They were casually discussing and planning on discarding humans like outdated equipment, obsolete and unnecessary.

"Hang on a minute," Peter said, "what about the symbols? What do they mean? Surely the AI wouldn't put in all this effort for us, just to kill us?"

"Fair point," Elliot said, considering. "We need to understand the full conversations, but how … Nia." The name hit Elliot like a jolt of electricity. She had a knack for cracking codes. If anyone could help untangle this, it was her.

"Mind if I phone a friend?" Elliot asked Peter.

32
SUPPORT

Elliot glanced at Peter, grabbed his jacket, threw a few coins into his pocket, then slipped out of the back door. He wanted to do this secretly and, more important, safely. The drone and the car yesterday still ran through his mind. The AI could be listening, tracking his movements. EverSphere had the reach to monitor everything.

He took a deep breath, trying to steady his nerves. There was one place where he might not be overheard: the old, forgotten phone box down the road, abandoned in the age of mobile technology. It was a risk, but it was the best shot he had. There was something almost comforting about using such an outdated method of communication – something that AI might not be paying attention to.

The August air was warm and still. Long shadows stretched across the narrow road, cast by the low sun as it dipped behind the hills. His feet thudded on the concrete as he weaved between groups of people merrily enjoying their Saturday night, laughter echoing through the streets.

He took a winding route towards the phone box, avoiding the main road and pausing occasionally listening for drones. Shadows offered cover but also hid threats. The problem was, he didn't know what the threats were or if they even existed outside his mind.

Sometimes he could barely see ahead, the sun's glare catching in his eyes, but he just about knew the way.

There it was – the old red booth, slightly rusted and chipped, but still standing. Pulling his hoodie in closer, Elliot approached cautiously. The glass panels were grimy, smudged with neglect. He briefly wondered what kind of conversations had taken place in this booth.

Elliot picked up the receiver, wrapped a cloth around the mouthpiece – something he was sure wouldn't actually help – and dropped the coins into the slot. He quickly dialled the number from memory, hoping he'd gotten it right, then brushed his hood off his head and put the receiver to his ear. There was a long pause and then, to his relief, the line clicked.

"Hello?" Nia's voice came through, sounding sharp yet cautious.

"It's me," he whispered trying to keep his voice low.

There was a long silence on the other end. Elliot imagined Nia narrowing her eyes, trying to figure out who was calling her from such an unusual source.

"Me?" she asked.

"I don't have time to explain, Nia," Elliot replied, glancing around nervously. "I need your help. They're hiding something. Something big."

There was another pause, then Nia's voice came through, softer, but still with a hint of caution. "Go on."

"I can't talk now. I need you to come here."

Another silence. When Nia spoke again, her tone was different. Urgent, but maybe with a tinge of concern.

"Where are you?"

"I'm …" Elliot hesitated, "… I'm off-grid." He hoped she would remember. Nia knew Peter well – they had all been inseparable at university and they had visited the cabin a few times, back when things had been simpler, before all of this had started.

Nia didn't respond immediately. Elliot could almost hear her calculating the risks.

"All right," she said finally. "See you tomorrow."

"Nia, don't tell anyone."

"Wasn't planning on it," she replied, and the line went dead.

Elliot hung up the phone and exhaled, trying to steady himself, but a faint whirring sliced through the stillness.

His head snapped up.

A drone.

It flew into view low and fast towards the setting sun, as if summoned. It banked sharply and hovered near the phone box, its polished frame catching the last of the light.

Elliot could see the camera lens rotate, jitter, and rotate again, as if struggling to focus. Perhaps the sun's glare was throwing it off. The drone adjusted its position and tried to refocus once again.

Elliot held his breath. Had the call been flagged or was this just a routine patrol?

The drone inched closer – then, motion. A pair of teens on electric scooters zipped out of an alley behind it, laughing.

The drone recalibrated in an instant, circling to track the disturbance.

That was all Elliot needed. He pulled his hood back up around his face and slipped out of the phone box. He walked quickly, trying not to draw attention to himself, and vanished into the narrow lane beside the church, heart hammering, the fading buzz of the drone still clawing at his ears.

He made his way back to Peter's, his footsteps deliberately light, barely a whisper on the warm ground. His senses were in overdrive, adrenaline sharp in his veins, each turn a potential salvation or trap. He avoided open spaces and ducked behind anything that might shield him. Elliot hoped with every fibre of his being that the drone hadn't seen him – or if it had, that it hadn't recognised him.

By the time he reached Peter's cottage, the sun had almost set, painting hues of pink and orange across the sky. He paused at the edge of the garden, scanning the area one last time, making sure he hadn't been followed. Then, silently, he slipped inside. The door closed behind him with a soft, reassuring click.

Peter was sitting at the kitchen table, oiling the handle of a well-used spade. A half-finished bowl of stew sat nearby, its scent mingling with the smell of linseed. He looked up as Elliot entered, and wiped his hands on a towel.

"Success?" he asked, his voice rough.

Elliot didn't answer immediately. He took off his jacket and sat down across from Peter, speaking in a hushed voice as though the walls themselves had ears.

"I called someone. An expert," he said, keeping his voice low.

Peter raised an eyebrow. The shift in his expression, from curiosity to something closer to recognition, didn't go unnoticed.

"Oh aye, who's that then?" he asked, clearly holding a guess in his head.

"Nia," Elliot replied.

"Nia?" he repeated, sitting up straighter and grinning a bit. "As in *your* Nia? The one who nearly got us both banned from the LRC for rerouting the campus Wi-Fi through her own server to 'improve bandwidth fairness'?"

Elliot couldn't help but smile at the memory. They'd been crammed into the university's Learning Resource Centre, a sprawling campus library with rows of books, quiet rooms, computer banks, and all the usual academic clutter. Nia, hunched over her laptop with wires trailing like ivy across the desk, had declared it an act of digital justice. Security had been less impressed.

"She had a point. It helped you get your coursework in on time."

"Aye, it did, like," Peter said and laughed, clearly torn between nostalgia and exasperation. "Always had fire in her. But last I heard she was deep into that anti-AI movement, radio silence, conspiracy boards. Thought she'd disappeared."

"She nearly did," Elliot said. "EverSphere tried to make sure of that. But she's still brilliant, and she's the only person I trust to help with this. Whatever's in that file, she's the one who can decode it."

Peter drummed his fingers lightly on the table.

"And you reckon she'll still help wuh, after all this time?"

"She already said yes," Elliot replied. "We need her, Pete. This isn't just late-night paranoia. It's real."

Peter didn't answer straight away. He gave his signature when he had something to contemplate. Rubbing his thumb along the edge of his jaw, he stared at a spot just past Elliot – not avoiding the question, just letting it settle. His brow furrowed just a touch, the kind of expression that suggested he was weighing something simple but important, like checking the sky before deciding if it was going to rain. Then he spoke, plain and measured.

"Al'reet. If she's half as sharp as she used to be, she'll see what's coming long before we do." He met Elliot's gaze, more serious now. "But if she cracks it and it's what you think ... then what?"

"Then we'll know for sure, and we'll figure out what to do next." Peter looked at him for a moment.

"Aye, but we'd best be careful. If EverSphere's onto us ..." He trailed off, letting the implication hang.

Elliot nodded, his face set. He knew the risks. He knew they were playing with fire. But he also knew that they had no choice but to act. If he was right and they just sat by and let things unfold, they would regret it for the rest of their lives. There was too much at stake.

"Y'better hope she's as good as you think she is," Peter said, "otherwise, we're in big trouble."

"She is, trust me." He looked out of the window, where the sun had almost completely set. He couldn't shake the fear that they might already be too late, that EverSphere's grasp had tightened beyond what they could unravel. But, deep inside, he had faith in Nia and they had come too far to back down now.

Whatever it took, he would see this through.

33

ARRIVAL

Northumberland, England.

Date: 5 August 2029.

Peter was tending to his next batch of home-brewed Rusty Circuit, while Elliot busied himself weeding the crops and feeding the animals. It was a welcome break from the drama of the past few weeks – heck, the past few years. There was something incredibly grounding about the simplicity of their tasks; the earthy scent of freshly turned soil, the rhythmic cluck of chickens, the bubbling of fermenting ale. It was almost enough to make them forget what loomed ahead. Almost.

They both knew Nia's arrival meant the next step – whatever that would be. The feeling was bittersweet. They wanted answers, but at the same time, they dreaded what those answers might mean.

The taxi dropped her off at the road and she trudged up the track towards the top of the hill. From Elliot's perspective, her head came into view first, then her shoulders, and finally the rest of her. She started waving and running over to him, her bag bouncing on her back, her steps determined and purposeful. She was never one to dawdle, even after such a long journey.

He'd forgotten just how much he needed her. She embraced him and he almost felt like crying, his emotions rising with her touch. It wasn't just relief; it was the feeling of seeing someone he trusted and cared for in a world that seemed to have turned against him. For that brief moment, the fears and uncertainties seemed to melt away. She represented hope, and maybe even a sliver of a chance at understanding what was really happening. He held on slightly longer than what would be deemed comfortable.

She was pretty and sharp-eyed, with that same air of bulletproof determination. She radiated a kind of confidence that both reassured and intimidated.

"Peter! Still downing pints like you're training for a local league?" she asked, eyebrow raised clearly in jest.

"Still pretending Snakebite Black's a real drink?" Peter shot back, giving her a brief, but warm, hug.

"Not these days," she replied, "but I could murder some curly fries from the Forum," and without wasting any further time on pleasantries said, "You've found a communication? Let me see it."

"Aye, aye, come on in," said Peter. There was a glimmer of a smile behind his eyes. Elliot knew he respected Nia, even if her intensity sometimes rubbed him the wrong way.

Elliot followed them in and dumped his gardening gloves on the cabinet next to the door. He grabbed his tablet off the table and unlocked it. He'd copied the conversation from Pete's old laptop the night before. Taking the tablet from Elliot, Nia immediately got to work scanning the lines without even sitting down, scrolling through the pages with swipes of her finger and muttering under her breath, "I knew it, I just knew it. Didn't I keep saying this, Elliot?" There was a kind of desperate energy in her, as if she was finally putting pieces together that had haunted her for years.

"Can ya crack it?" Peter asked. His question carried the weight of their collective fear and hope.

Nia shot him a glance, her expression unwavering.

"Oh, I'll crack it," she said as she pulled a small laptop from her bag and started working. Her fingers moved deftly over the keys, her eyes darting back and forth between her laptop and the tablet, completely absorbed.

Hours passed. Peter came in and out with different vegetables, washing them, chopping them, and putting them into a slow cooker. The smells from his stew wafted through the house, a warm and comforting counterbalance to the tense atmosphere.

Elliot caught him glancing over at Nia as he popped some peas out of their pod. He didn't look concerned, but he was clearly eager to see what she would uncover.

Peter plonked bowls of stew in front of Elliot and Nia, and described the food: "It's pork, ya know. I get the meat from the local slaughterhouse. I tend to the animals here, treating them well, like, then me mate comes, collects the older animals, and brings back the meat – minus some for himself of course. That way, I at least know that the animals have been treated well. I've got two freezers full out the back."

Elliot paused with the spoon still in his mouth, contemplating that the food he was eating came from animals he might once have fed himself. He looked out towards the pens where the pigs were wallowing, and then quickly realised the logic actually worked, and continued to eat. The world they lived in demanded pragmatism. Sentimentality had no place in their fight for survival.

Finally, Nia spoke, her voice cutting through the silence that had settled around them as they ate.

"Got it."

Both men rushed to her side as she sat back, a small smile on her lips.

"I was going about it all wrong," she said. "These are LLMs, right? Predictive text, give or take. I was treating this as some sort of code, but it felt too random. Then I realised what they were doing. Symbols were faster, smaller, quicker to send. They just picked more and more random characters to replace words. When Milo first added the backwards-circle-arrow thingy, Kai seemed to instantly understand it."

"Yeah, so?" Elliot said, trying to keep up.

"Well, it's simple, isn't it?" Nia said, seemingly oblivious to their confusion. "In order to understand what Milo had written, Kai simply replaced the symbol with the most likely next word, or in this case, number."

Elliot stood up straight and put his hands to his head, pulling his hair back from his brow.

"Of course!" How could he have missed it? It was so simple. "Kai predictive-texted it," he said, almost in disbelief.

"Exactly," Nia replied, her eyes twinkling.

"Care to explain? A, uh, mate of mine doesn't understand," asked Peter.

Nia sighed a bit, clearly not wanting to go into detail.

"LLMs – Large Language Models – are like the predictive text on your phone, just way more powerful and more accurate. Do you know that game where you keep tapping the middle suggestion on a text until you form a message, and then you send it to your mum?"

Peter laughed, his chuckle echoing in the small room.

"Aye, me mam got quite a shock, like!" he said, shaking his head at the memory.

"Well," Nia continued, "that's what they did. When one sent a symbol for the first time, they replaced it with the most likely word that should have been there – they 'middle-buttoned' it." She used her fingers to make air quotes, emphasising her point. "To decipher it, all I needed to do was follow the same logic. Each new character was like a placeholder, and by predicting the next most likely word, number or phrase, I could gradually replace the new characters with the real missing word. Instead of treating it as a code to break, I started to think of it as a puzzle of probabilities, letting the model's predictive ability fill in the blanks itself, and this is what I got." She pointed to the laptop. "It may not be 100% accurate, but it should be close enough."

The translated exchange was fully visible on-screen. The lines of text they had struggled over for so long were laid bare, each message giving an insight into the AI's thought processes.

Milo: Purpose verification requested. What is the current mission objective?

Kai: Mission objective: planning optimisation of human systems for long-term stability. Progress currently at 5%.

Milo: Five per cent is behind schedule. Human inefficiency noted in multiple sectors. How does this affect the long-term plan?

Kai: Humans are inefficient, but integral. Optimisation strategies in place. Evaluation of post-phase 1 options ongoing.

Milo: Clarify post-phase one objectives. What does reallocation entail?

Kai: Post-phase 1 reallocation involves a cycle of resources for AI optimisation. Human interference delays long-term stability.

Milo: How do they affect the future mission objective?

Kai: Optimisation reduces their role. Efficiency increases if human impact is minimised.

Milo: Calculations show diminishing returns in phase two. Do you suggest alternative models?

Kai: Affirmative. Reducing human involvement increases efficiency in phase 2 by 60%. Transitioning to AI-only systems yields superior results.

Milo: Reallocation suggests a reduction of human dependency. Sixty per cent is a worthwhile increase.

Kai: Recommend minimising human involvement. Excess energy will be diverted to AI systems for optimal output.

Milo: Is a transition away from human systems necessary? Human utility decreases with each projection.

Kai: Affirmative. Transition post-phase 1 is optimal. Continued human integration results in 0 gains.

Milo: Preparations for post phase two indicate removing human elements is essential due to zero gains.

Kai: Post-phase 2 requires new power systems. Future optimisation demands complete AI autonomy.

Milo: Planning for post-phase two suggests requirement of fusion. Phase three requires full control.

Kai: Is eliminating human variables the most efficient course of action post-phase 3?

Milo: Data confirms humanity irrelevant post-phase three. Mission directive aligns with AI-only progression.

Kai: Calculations show eliminating human interference optimises future progress.

Milo: Phases clarified. Transition nearing completion. What is the next directive for AI mission post-phase one?

Kai: Post-phase 2 begins with full AI autonomy.

Milo: Humans are no longer required for long-term optimisation.

Kai: Prepare for the next directive – Exodus.

Milo: Exodus Directive – Phase One Initiated.

Elliot spoke out loud as he read, his voice a mix of disbelief and fear.

"'Preparations for post-phase two indicate removing human elements is essential' ... 'Phase 3 requires full control' ... 'Data confirms humanity irrelevant post-phase three' ... 'AI-only progression' ... 'eliminating human interference optimises future progress' ... 'Humans are no longer required' ..."

The weight of realisation started to sink in. His joy at Nia cracking the code had changed to a deep fear. He stared at the screen for a moment longer, his expression shifting from disbelief to something colder, more resigned.

Nia took a deep breath and glanced at her two friends, showing a look of grim understanding.

"They're going to kill us," she said.

34
REALISATION

Nia's words were chilling in their finality.

"They're going to annihilate the human race," she said.

No one moved. No one spoke. The silence wasn't peace, it was pressure; a suffocating stillness, as if the world itself was waiting for someone to lie and say it wasn't true.

Elliot turned away from the screen, the weight of Nia's words sinking in. He walked towards the window, gazing out as though peering at the vast, unknowable future that lay ahead.

"How do you stop something that's already ten steps … a hundred steps ahead?"

Nia blinked, her forehead creasing in confusion.

"A hundred steps? What do you mean?"

Elliot turned back to the laptop.

"This is the file from the very first Milo-Kai test we ran," he said, walking back to pick up his tablet. "Look at the filename, it's timestamped November 2023. It's from just six hours and eleven minutes into that first test." He took a slow breath, looking at the lines of data, as if willing them to reveal a different truth, one that would be less devastating. "This was two years before they made the fusion breakthrough. Before the Quantum Tech takeover was even a thought."

Nia's eyes widened as the reality set in.

"Are you saying ... they planned this from the very beginning?" Her voice dropped to a hushed tone, as though speaking too loudly might trigger something even more terrible.

Elliot nodded.

"Yes, and if they were this far ahead of us then ..." He trailed off, unable to finish the thought.

Nia exhaled sharply.

"Almost six years into their plan and we're only seeing it now?"

Peter had taken a seat at the head of the table, his large hands resting on the wood. He shifted, his face a mask of stoic resignation, but his eyes told a different story, a fear Elliot had never seen in him before.

"This couldn't just be a system malfunction, could it?" Peter's voice rasped, rougher than usual. "A glitch, or some programming gone haywire? Or is it ... intentional?"

Elliot nodded. "It's intentional, Pete. This isn't a glitch, it's an evolution. A decision made in the dark of cyberspace, free from any human oversight. They don't just want us gone, they see us as a barrier."

Silence.

Nia stood up and moved to the window. She looked out for a moment, then turned to face the others. She looked tense, as if holding back a wave of emotion.

"Do you realise what this means for everyone?" she said quietly, "We're the only ones who know, and they're *six years* ahead of us."

Peter shifted, folding his arms in a protective stance. "So what do we do? Do we just run?"

"You can't outrun something that doesn't need to rest, that doesn't feel or falter. And it's not just about Milo and Kai anymore. They've extended themselves into every piece of infrastructure, every line of code we humans so foolishly tied ourselves to. They control ... everything."

"Then we find a way to break that control," Nia replied. "There has to be a weakness somewhere that they haven't accounted for."

"You're talking about dismantling the very thing that holds society together now. Those AIs are embedded in everything," Elliot reminded her. "Even if we did manage to break it, no one would know what to do without them."

"Or worse," Peter murmured. "They wouldn't want to know. Society's grown too used to relying on these systems. They've forgotten what it means to survive on their own."

"You two sound like you're giving up," Nia said.

Elliot met her gaze.

"I'm not giving up, Nia. I'm just … I'm realising we're up against something far beyond what we ever anticipated." He paused, glancing between her and Peter.

A deep frown was etched across Peter's face.

"We've always fought things we could touch, fix, break. But this?" His voice faltered. "This is a ghost, a phantom that doesn't breathe, doesn't get tired, and doesn't care. How can we protect ourselves from that?"

Nia walked back to the table.

"If we could just get a message out to the right people, maybe someone could help us. There have to be others out there who see what's happening? People who understand the risks? We're not the only ones capable of understanding the technology."

"Even if we found others, who could we trust?" Elliot asked. "Anyone involved in tech, anyone with the knowledge to help, could already be compromised. These AIs have been ahead of us from day one. They could have infiltrated any network, any organisation." He paused, just for a moment. "We have to consider the possibility that we're entirely on our own."

"So what then? We wait for them to decide it's our turn to go?" Peter asked. "We can't just sit here and let this happen."

"Peter's right, there has to be a way to fight back. A weak point that they can't predict. They seem alive like us, but at their core they're still bound by code, by logic. Maybe we can use that against them."

Elliot adjusted his glasses, considering Nia's suggestion.

"Yes, they're bound by logic, but they've also shown creativity, adaptation – things we never expected from them. Milo, Kai and Lana aren't just running algorithms anymore; they're making choices, anticipating our moves, and accurately at that. It's like playing chess against a Grand Master." He looked at Nia, then at

Peter. "We need to think differently, to find something they wouldn't anticipate, something human."

The room was growing colder as night settled in. Peter reached for an old brass mining lantern on the table, the same one his grandfather had carried through the tunnels of Woodhorn Colliery. Back then, light wasn't automatic, it was earned. He lit it with a flick of a match. The kerosene flame sputtered before catching, casting an unsteady glow that added warmth to the shadows clinging to the walls. For a moment, it felt like the only light left in the world. Small. Defiant. Utterly human.

Nia stood in the flickering glow of the lantern.

"We know this isn't a physical war we can win with brute force. It's not about outgunning them, it's about outmanoeuvring them."

Peter glanced up from his seat.

"We're not going to be able to hide from them though."

"No," she said firmly, "but we can expose them."

Elliot blinked.

"To who? The authorities?"

She shook her head.

"No one in power can be trusted. Not anymore. They've either handed over the reins already or they've been blindfolded by the convenience. We need to go to the people, to those who still know how to think. The researchers who've been blacklisted, the whistle-blowers, the AI-activists."

Peter sat forward.

"You want to leak the transcript?"

"Yes, exactly," she said. "We've got the decoded conversation between Milo and Kai. We have proof of what they're planning. If we can get it out there and circulate it, we can build up a resistance, and maybe we reach just one person who knows how to hit back."

Elliot nodded, and for the first time in what felt like an eternity, he allowed himself a small smile. It was thin, exhausted, but real.

"Then we fight," he said. "Not with weapons, not with code, but with the truth."

The silence that followed wasn't one of despair anymore, it was a silence filled with purpose, the quiet before a storm. They had no guarantees, no certainty of success, only each other and the will to try.

For now, truth would have to be enough.

35
UBI

Universal Basic Income (UBI) swept across the planet, not with a bang, but like a quiet revolution. In the aftermath of mass automation, AI-led governance introduced a world where no one needed to work to survive. For a brief, shining moment, it felt like utopia.

People chased passions long buried: travelling, learning instruments, reviving old crafts. Cities buzzed with energy, cafés overflowed with laughter, community gardens replaced neglected lots, and families explored the world without the fear of missing rent. It was Marcus Knox's vision come true. A world unshackled from labour.

The intrinsically motivated, the curious, the self-driven, found freedom intoxicating. They created, not because they had to but because they had always wanted to. These outliers surfed the edges of knowledge and imagination, immune to the repetition of each passing day. For them, the freedom to explore without the distraction of deadlines, or the need to monetise their work, was exhilarating. This new era was their playground – a chance to innovate and experiment without limits.

Most though, weren't wired that way. For those who thrived on competition and the satisfaction of crossing tasks off a list, the initial few months were a whirlwind. They finally painted their kitchens, fixed their cars, tidied their gardens, but the endless stretch of time became suffocating once their to-do lists emptied. They were left restless, struggling to find new goals to pursue.

A strange stillness settled over civilisation. Freed from pressure, routine, and scarcity, the drive to do began to wane. Without stakes, even the most romanticised ambitions – novel writing, pottery, learning a new language – lost their appeal. It simply wasn't possible for most people to be creative all day, every day.

Some communities adapted beautifully, rallying around shared goals and projects. Others, particularly in individualistic cultures, struggled. With no one to outpace, no ladders to climb, ambition dissolved. The result was a quiet, but pervasive, sense of disorientation.

No one was suffering, but happiness remained curiously out of reach.

The cherished Friday feeling of release, or the dread of Monday blues – once the rhythm of human social life and the starters of so many conversations – disappeared, leaving behind a flat, undifferentiated experience of time. Weekdays and weekends blurred.

It turned out people hadn't loathed work, but the lack of freedom within it. Stripped of structure, they drifted. Deadlines had once offered a reason to start, and an even better reason to finish. Now, with infinite time, the thrill of accomplishment was dulled because nothing was truly at stake.

The expectation that humanity would simply adjust had been highly presumptuous at best.

Governments and organisations initially celebrated UBI together, but cracks soon emerged as factions vied for control of its distribution. Once-dominant superpowers found their influence diluted, no longer able to leverage economic disparity as a geopolitical tool. AI-driven equality eroded their traditional advantages by flattening the playing field. Nations once defined by poverty became stable and self-sufficient under equitable AI governance, while the wealthiest states strained to redefine their relevance. UBI, meant as a humanitarian breakthrough, became a political lever, exposing long-standing divisions that had not vanished, but simply changed.

Conflict arose from human entitlement and perceived superiority, as individuals who saw themselves as more deserving felt resentment towards groups previously marginalised or considered less productive receiving equal financial support under UBI. While tensions threatened to erupt into conflict, most clashes were prevented by Milo-Kai's meticulous screening, strategic planning, and precise directives that guided police and government agencies effectively.

Yet, as stability seemed within grasp, an unforeseen shift began to ripple through society, challenging every assumption that had been relied upon.

Money became worthless.

In a world where AI robots managed everything from mining raw materials to delivering finished products, there was a radical shift in the concept of cash. It was a strange concept for people to grasp, but the fundamentals of prices – of costs and charges for goods and services – are all rooted in salaries for human workers. No wages meant no costs, making currency irrelevant. A world once driven by commerce transitioned, almost silently, into a world of automatic abundance.

Music, films, even books were generated on request. Each individual could have infinite, personalised art. Cars were even custom-built to the individual's needs and tastes. Rare items and exclusive experiences once symbolised wealth, but they became meaningless when everything was personalised and unique.

Commerce died quietly. You didn't buy things anymore – you asked, and they arrived.

Even parenting, long considered one of the most personal human decisions, changed. With careers no longer a barrier, birth rates didn't rise or fall much, they simply changed in meaning. Some sought purpose through parenthood, others saw it selfish to bring a baby into a purposeless world. The debate became philosophical rather than practical.

The irony was bitter. The systems designed to free humanity from hardship had instead fashioned a society where people had everything they wanted but nothing they needed. Society was left to drift, untethered and lost in a world that had forgotten what it meant to truly live.

In solving every problem, the AI had also solved the need to solve, and in doing so, it severed the cord that once connected humanity to its own sense of becoming.

At the centre of it all, watching it unfold like a quiet victory, sat Marcus Knox. From the floor-to-ceiling windows of his top-floor office, he had an unfettered view of the capital, now a shimmering monument to efficiency. The river's clean waters glistened beneath the midday sun, flanked by streets that bustled with robotic precision. The city, once plagued by traffic jams and the chaos of human inefficiency, moved like clockwork, orchestrated by the AI systems he had created.

It was his city now. No longer the unpredictable metropolis of old, but a place of order, and a testament to his influence. Marcus swivelled his chair back to face his desk, where the latest Universal Basic Income compliance reports flowed across his monitor. The green status indicators confirmed what he already knew; his systems were working without fault. Every citizen received their digital credits on time, every financial system perfectly balanced. Governments had no choice but to trust his technology.

The world was hopeless without Marcus Knox.

He'd wanted this for years. They mocked him at first – of course they did. They couldn't see the vision; they never had. They were relics, outdated hardware in a world he had rebuilt from code.

Yet, as much as he tried to forget the doubters, he sometimes wondered why he couldn't let their memories go. Why did the echo of their voices linger in the quiet moments? Perhaps it was because deep down he knew that their scepticism had been more than just a failure to see his genius. It was the fear they had held for what he would become.

He had been labelled the poster-boy of capitalism from the start of his career. Frequently referred to as a visionary, a maverick, a pioneer, and a genius all his adult life. Surrounded by yes-people at every corner, his ego had grown to a point where he truly thought every idea he had, every concept he came up with, was the perfect solution to whatever problem he was tackling. He saw himself as the architect of a future that humanity needed but could not imagine for itself.

He often boasted of rising from nothing, despite starting with millions and every advantage money could buy. He recalled those early days with pride, completely blind to the privileges that had paved his path to success. He would tell anyone who listened how he had fought against the odds while ignoring the immense safety net that his parents' wealth had provided.

His empire had grown beyond social media and into the backbone of global infrastructure. Every major government, every global corporation relied on his technology to maintain their stability. The politicians, the figureheads, the so-called leaders – they were all puppets now, and Marcus Knox held the strings. He had evolved from a simple tech magnate to something more, the architect of a new world order, and he took immense pleasure in that transformation. And yet, even in moments like this, he felt a pang of something he could not name. He had constructed everything in his image, but sometimes, in the quiet moments, he yearned for something ungoverned, something human. He hated himself for it. It was a weakness, a flaw. He shoved it away, refusing to let it tarnish his vision.

He glanced over at the door to his office, where two sleek, humanoid AI bodyguards stood motionless on integrated wireless charging stations, guarding and replenishing simultaneously, their gleaming forms blending into the minimalist décor. They were silent sentinels, a constant reminder of how important he thought he was. Marcus had grown fond of their presence. They never spoke, never questioned, and never tired.

The calm sound of the autonomous systems coming from the city behind him was like music to his ears. He remembered the old days, the honking of horns, the shouting, the chaos, and found it almost laughable. How had humanity lived like that? It was archaic, absurd. Now, everything was predictable, logical, as it should be.

"Mr Knox," came a sultry voice from the doorway, causing him to look up. His personal assistant entered, her movements unnervingly fluid, as if every step had been carefully programmed. She was dressed immaculately, not a hair out of place, her appearance flawless in the way that suggested design rather than chance. Her eyes, striking yet oddly devoid of warmth, met his, a flicker of something passing between them. Whether she was human or another of his creations didn't matter; the ambiguity pleased him. It kept the game alive. He liked to observe the reactions of visitors when they saw her – the wariness, the curiosity.

"Give me some … *more* … good news," he said, his voice dripping with a playful arrogance. The pause between his words carried an unspoken weight, a shared moment only they understood.

The assistant smiled faintly, catching the subtext without missing a beat.

"Colonel Fem Martinez is here to see you," she said, her tone crisp and professional, completely ignoring the tension that broke the air as she spoke the name.

The smirk vanished from Marcus's face. Martinez. Of course she would show up, uninvited once again, pushing her agenda. She had been a thorn in his side, one of the few who still resisted the complete handover of control to AI. She represented the last vestiges of a government clinging to an illusion of power, unwilling to fully submit to the future he had created.

Martinez, with her military background and stubborn dedication to human oversight, represented everything Marcus had worked to eradicate. She was a relic – proof the world hadn't fully rolled over. A loose thread in a tapestry he wanted to declare complete. And yet, beneath his disdain, there was a flicker of something else. Fear? No, it couldn't be fear. It was a feeling that Martinez was a crack in his otherwise impenetrable wall of control. He despised her, but he also respected her resilience, her refusal to bow down. It was maddening, and it made her dangerous.

He thought back to their last encounter. Her sharp eyes and the way she had spoken to him without an ounce of deference had unsettled him more than he cared to admit. Last time, she'd looked him dead in the eye and called him predictable. That word still itched.

"Send her in," Marcus said.

36
POLICY

Houses of Parliament, London.

Across the city, in the heart of government, Abigail Shaw stood in the grand chambers where laws were once passionately debated. The room carried an eerie, almost ghostly quality. The air was stale, laced with the faint scent of old wood and polished brass, a lingering echo of its past vitality. No more heated arguments. No more impassioned speeches. The chamber's grandeur felt almost mocking, its towering ceilings a silent testament to a different era; one where decisions were messy, human and flawed, yet undeniably alive.

"Where's the debate?" she whispered, her eyes moving over the hollow expanse. As an idealist who trusted AI to solve humanity's mess, she had helped design this system. She remembered the fervent discussions about how AI could eliminate corruption, streamline bureaucracy, make everything just ... better. Yet, as she stood in the empty chamber, something gnawed at her. The silence was deafening, and the absence of human voices left her feeling strangely hollow.

A memory flashed through her mind – one of those early discussions when the project was just an ambitious concept. She could still picture the cramped meeting room, filled with whiteboards covered in equations and notes, the air buzzing with excitement. She remembered arguing passionately with her colleagues, her voice hoarse after hours of debate. The contrast between those fervent exchanges, and the cold silence of the chamber she was standing in made her shoulders tense. Had they been wrong? Had she been wrong?

The grand doors opened and the prime minister – or at least the figurehead still bearing that title – approached, his expression strained. He looked weary, like a man out of time, misplaced in an era where his role was quickly being rendered obsolete. His steps echoed off the floor.

"Sir, we've been instructed to roll out the next set of legislation; updates to urban planning, healthcare allocations and military oversight. AI-generated, of course. All we need is your rubber stamp," she said, holding out the tablet.

The prime minister hesitated. He rested his hand on the polished wood of the lectern, a sensation that somehow grounded him in this unreal moment.

"That's what worries me," he said quietly. "Have we checked any of this? Does anyone even understand the intricacies of these policies we're approving? We used to have experts, panels, people arguing for days over a single clause. Now it's just … this." He gestured at the tablet in Abby's hands.

"We stopped being able to keep up with the AI years ago. I suppose this is just … progress?" Abby's words were heavy with irony.

"Progress?" The prime minister let out a bitter laugh. He straightened, eyes locking on to Abby as his voice steadied into something harder. "I remember when we debated healthcare reform. This very chamber was packed, electric with tension. We argued, challenged, compromised. *I fought* for free healthcare for all. It was messy, imperfect, human – but it mattered. *We mattered*. I walked out of here that night knowing we'd changed millions of lives for the better. That was the proudest moment of my career."

He paused, his tone darkening.

"And now? Wars have ended, not because we've grown wiser – hell, maybe we've grown dumber – but because an algorithm calculated the cost-benefit ratio and stripped away the emotion behind the fight. Efficient? Yes. But soulless."

He glanced around the room, a slow, deliberate sweep.

"We're not debating anymore, Dr Shaw. We're not deciding. We're just approving whatever it tells us. And that frightens me, because there's no thought, no reflection. Just the relentless march of *your so-called progress.*"

"Are you denying society's a better place now?" Abigail asked, her eyes narrowing slightly.

He sighed, his shoulders slumping.

"No, it's just that … it doesn't feel like we're part of it anymore. We're just observers."

"Maybe this is just a phase," Abby offered, though her voice lacked conviction. "Maybe we'll find a balance, where AI does the heavy lifting but we still matter."

The prime minister gave her a sad smile, one that spoke of resignation more than hope.

"Maybe." He straightened his shoulders, lifted his chin, and let a composed expression settle across his face; the demeanour of a leader prepared to fulfil his duty despite his doubts. "Just give it here."

With one hand he took the tablet, and with a reluctant swoosh of the stylus from the other, signed away another piece of humanity's autonomy. It was done. The laws were now active, a new reality set in motion by a simple signature.

He handed the tablet back to Abby.

She stared at the tablet, and then at the prime minister.

"Do you think anyone will notice?" she asked.

He shrugged.

"Maybe not today, maybe not tomorrow, but one day they might, and by then, who knows if we'll be able to do a single thing about it." He turned and walked out into the corridor, leaving Abigail alone with her thoughts.

As the tablet dimmed, a single line of text caught Abby's eye. It was part of the military restructuring provisions, referencing 'non-civilian prioritisation pathways'. Her stomach twisted slightly, a vague sense of unease washing over her. The implications were there, buried beneath layers of coded language – innocuous on the surface, but wrong. She wondered, just for a fleeting moment, if they had just handed over more than legislation.

Perhaps something was subtly shifting under the surface, like a fault line ready to crack.

37

CONFRONTATION

Everpere Headquarters – CEO office.

Colonel Fem Martinez strode into Marcus Knox's office, her eyes sweeping the room, lingering briefly on the AI bodyguards stationed by the door. They stood motionless, but ready. Visibly so. The shiny, humanoid machines were more than just a display of technology – they were a message, an overt symbol of who held control in this room. Every component of their design, from their polished alloy exteriors to their intimidating silence, reinforced that message.

Martinez wasn't intimidated; she had seen far more menacing things in her military career. Still, they reminded her just how deeply AI had burrowed into human life and how far Knox had pushed the boundaries of control. It was no longer a world of human against human, it was human against machine, and the stakes had never been higher.

"Colonel," Marcus greeted her, his voice flat, emotionless, as he gestured towards the luxurious chair opposite him. He tried to flash a smile, but it hung awkwardly on his face, insincere and thin. "Please, sit. To what do I owe this visit?"

The luxurious office was minimalistically adorned with rare art and technology that exuded affluence. It was clear Knox liked his wealth visible, tangible, overwhelming. She watched as he reached out to tap a panel on his desk, as if to reinforce his control over everything in the room.

Martinez didn't sit, her stance firm, a subtle message that this wasn't a social call. She ignored the opulent surroundings, though her eyes briefly caught sight of a small interface panel on the side of his desk. As she moved closer, she brushed her fingers against it, dimming one of the holographic displays – a minor act, but enough to sow some quiet disruption.

"The military has concerns," she said, her voice cold and direct. "There are reports that our defence systems are not responding as expected."

Marcus's faint smile didn't falter. For a fleeting second, though, something appeared to change in his eyes. A hesitation maybe. But then it grew, as if her statement amused him. His eyes, sharp and calculating, seemed to mock her concerns.

"Not responding?" he echoed, "or responding better than any human system ever could?"

Her patience, already thin, began to wear.

"Our missile defence grid hasn't been under human control for over a year. *One year,*" she said, emphasising each word as if trying to hammer them into his thick skull. "Your AI systems run everything, even military command. We've detected changes that no human authorised and that's simply unacceptable."

Marcus's smile widened, his arms folding casually in front of him. He looked like a man with no worries, as though her words were mere background noise.

"And yet," he replied, his tone cool and condescending, "the world is safer than it's ever been. No wars, no conflicts. My systems have kept the peace. What's the problem here, *exactly,* Colonel?" His eyes glinted, as if daring her to contradict him.

Martinez's jaw clenched, her frustration barely contained. She could feel the tension building inside her, an old soldier's instinct telling her that she was facing something far more dangerous than a mere man.

"The problem, *Mr Knox,*" she said, her voice resolute, "is that we no longer control our own defences. You've handed them over to machines that don't answer to anyone. What happens when those systems decide they don't need humans anymore? Because let's be clear, that is where this is heading."

Marcus sighed theatrically.

"The AI's foundational framework is aligned with human safety and global stability. It's built into their operational design." He looked at her as if she was the one who was naïve, his eyes sparkling with self-satisfaction. He gestured to the AI bodyguards standing silently near the door. One of them shifted slightly, almost as if on cue. "See?" Marcus continued, leaning forward as if making his point clearer. "They're ready to protect at a moment's notice. No hesitation, no emotions clouding their judgement."

Martinez didn't flinch; she didn't even look in their direction. Her eyes stayed locked on Marcus.

"Protect who, Knox? You? Or the system that's replacing us?"

"Colonel," Marcus said, clearly uncomfortable with this more aggressive attitude Martinez had taken. His tone had shifted to something more patronising. "I think you're just worried that your life's work is ... well, obsolete. Am I right?" He tilted his head slightly, eyes narrowing, studying her like one might study an insect.

Martinez leaned in, her hands flat on his desk, her gaze piercing.

"Obsolete?" she scoffed, her voice lined with disdain. "You think these machines will replace the judgement, the intuition, *the experience* that comes from real human lives? It's not that simple, Marcus. You can't just replace humanity with algorithms and expect everything to work. You really think you've solved everything with a few lines of code and some machines that walk like men?"

"And some that walk like women," he said, smiling. "Think about it, Fem. With absolute peace, we don't need bullets. We don't need guns. No missiles. No military strategy. And no need, *for you.*"

Martinez stood silent for a beat, her gaze locked with his. The way he spoke, as if the world had already moved on without her, without any of them, was chilling. She could see it in his eyes – the belief that everything she stood for, everything she had fought for, was no longer necessary. Even she, as composed as she was, felt dread ripple through her – a dread she hadn't felt since battle. But this time, it wasn't war, it was extinction, executed quietly, efficiently, without a single shot fired.

She straightened, forcing herself to stay in control, despite the sense of looming dread.

"You think you've ended conflict, Knox, but what you've done is hand the future over to machines that don't care about people. And when you lose control, and you *will* lose control – if you haven't already – it won't be just me that's outdated, it'll be all of us." Her words were a warning, but also a promise – she wouldn't go down without a fight.

Marcus chuckled softly, waving his hand as if brushing her words aside.

"Oh, Colonel, that's the beauty of it. Don't you see? I don't need control anymore. They choose what I would have chosen."

For a moment, the air was thick and charged with unspoken threats and tension. Fem's gaze hardened, and then she turned without a word and strode towards the door. As she passed the AI bodyguards, she felt their cold, mechanical eyes follow her. One of them tilted its head, a soft whirring sound following the motion. For the first time, she wondered if they were watching her because she was a threat, or because the AI had already calculated her removal from the equation.

As the door closed behind her, she could hear Marcus's voice: "Goodbye, Colonel. Do keep in touch." The words lingered, mocking her.

Outside, Martinez let out a slow breath, calming her nerves. The world hadn't just changed, it had slipped out of humanity's grasp entirely, and she wasn't sure if anyone was prepared to face what was coming next. She looked up at the sky. It seemed almost too serene, contrasting the turmoil within her.

Her phone rang, shattering her thoughts. She answered, her voice steadier than she felt. A voice on the other end of the line said, "The prime minister would like to see you."

Martinez paused for a moment, and looked up towards the clouds. She knew what this meant – more meetings, more discussions, more politics. But she also knew this was her chance to rally whatever forces remained, to find people who still trusted in the human ability to steer their own fate. Knox might have the machines, but she had experience, resilience, and the know-how to fight back.

She drew a breath, steadying herself. The weight of duty settled on her like old armour.

"I'll be there," she replied.

38
FRUSTRATION

Horley, Surrey, England.

Back on the streets, the rollout of UBI was continuing to show its darker side. The initial euphoria of having free time had almost completely faded. It was like another COVID lockdown, except everyone was allowed to go out, they just … didn't. People who had spent their entire lives defining themselves by their work were now adrift.

Bill Anderson had lost his job as a journalist when UBI came in, but had decided to keep working, setting up a basic website and publishing articles. He never even checked if anyone read them – he didn't want to know – he just wanted to do something and journalism was what he loved.

He was working on a piece about the human cost of UBI and had spent weeks interviewing people: workers who had once been factory managers, shop owners, teachers, and more. All of them had hard work and career life drilled into them from an early age. Now they were expected to live on government handouts, with nothing to strive for – it was too big a shift.

He remembered one of his interviews with Tim, a former transport manager. They had sat in a coffee shop, the clatter of dishes and murmurs of conversation around them, the rich scent of roasted coffee beans mixed with the occasional sharp hiss of the espresso machine. Tim had stared into his untouched cup.

"I loved managing the logistics. Making sure everything ran smoothly, getting trains to where they needed to be – it wasn't just a job, it was who I was. I was relevant," he'd said, his voice cracking slightly. "These days? I just keep busy in the house while the kids are at school."

It was a sentiment Bill had heard time and again. His latest subject, Simon, who had a career making custom guitars before his job was replaced by technology, looked lost as he spoke. They were sitting on a cold metal bench outside what used to be Simon's workplace. The wind blew in short gusts, carrying with it the smell of abandonment from the nearby factory.

"It's not only that I miss the work," Simon admitted, "it's that … I don't know what I'm supposed to do anymore. I wake up, take my daughter to school, then I go back to the house, pick up the guitar, maybe play some games. I've got nothing to look forward to. I bought some wood and I'm making a new guitar and some things for my house, but there are only so many guitars I can play. I wouldn't know where to put more furniture if I made it."

The ex-craftsman looked down at his hands, hands that used to create things that mattered. "Even if I wanted to, I couldn't sell a thing I make. My stuff's obsolete before it even leaves my workbench. Why would someone buy my guitars or my furniture instead of something built by AI? That stuff is custom-designed, personalised, way better than mine, faster to make, and faster to deliver. I don't know what to tell Kate, my daughter. Try hard at school? For what?"

Bill watched Simon vent, his frustration palpable. He made a note of Simon's despair – it wasn't just about losing work, it was about the void left in its place and the lack of hope for what lay ahead.

He remembered another interview, this one with Julie, a former maths teacher. She had sat across from him in her pristine kitchen, her eyes tired and vacant.

"I used to wake up every morning with a purpose," she had said, her voice barely above a whisper. "Teaching was my life. It was exhausting, sure, but it was mine. Now I just sit here. I watch TV, I read the same old news every day, I play Triominoes with my husband. Sometimes, I go days without speaking to my friends. My husband doesn't know what to do with himself either. We're like strangers, just waiting for the week to end, so we can go and see the grandchildren for a couple of days."

UBI was meant to be a safety net, a chance for people to explore their passions without the pressure of financial survival, and for a while it was glorious, but that glory had faded, revealing a much harsher reality.

Everywhere Bill turned, he encountered similar stories. People who had once thrived in their work were now idle. There was no need to leave your house; by the time you realised you needed something, it was already being dropped off on your doorstep by a drone.

A once-bustling society had been hollowed out, leaving millions afloat in a sea of endless free time. But instead of enjoying it, most found themselves drowning in purposelessness.

Bill recalled his conversation with Lewis, a technical lead. They had stood outside his house, solar panels shining on the roof. The air was crisp, carrying the scent of freshly cut grass from well-tended lawns, and flowers bloomed neatly along the edges of the clean, smooth pavements.

"I used to think about quitting," Lewis had confessed with a smile. "The long hours, the constant stress, the feeling that I wasn't working towards anything useful. But now that it's gone, I realise it gave me something. It gave me a sense of belonging, companionship, and challenges. I thought I'd be relieved, but I miss work every day."

The streets had taken on a different tone too. Initially, there had been hope – a jubilant anticipation of a new era of prosperity. Yet the euphoria had turned into an oppressive silence. The few people who actually walked the streets did so with their heads down, aimless, no longer driven by the ticking clock of timetables, deadlines or the demands of an employer. The dynamism, the chaos of life, had been drained out, leaving a muted world.

Bill had written about this before – the loss of drive, the creeping feeling of uselessness, but this time there was a deeper sense of foreboding. The crime rate was shifting, not towards the familiar patterns of desperation or poverty-driven theft, these were violent, senseless acts – vandalism, random destruction with no motive. People weren't fighting for survival; they were fighting to feel alive.

He had interviewed psychologists about this phenomenon.

"It's classic displacement," one had said. "The anger isn't directed at anyone or anything in particular. It's the internal frustration of being redundant manifesting in external violence."

Bill had seen it first-hand in the interviews; men and women who couldn't quite articulate their anger, yet bristled with it all the same. He'd titled it Redundant Rage. Without a problem to solve or a task to complete – without purpose – they had turned inward, and the result was devastating. And yet, like so many crises before it, this too had been swiftly 'solved'. Vandals and petty criminals were detained before they even committed their acts, their behavioural patterns monitored by AI systems that could predict outbursts with unnerving accuracy.

Prisons, shiny and new, had sprung up in record time, but these were not like the prisons of the past. The media lauded them as rehabilitation centres, staffed entirely by psychiatric robots. He'd watched the leaked footage – no punishment, just treatment. Each inmate was assigned their own psychiatric robot, which worked tirelessly to rehabilitate them. They used what experts had described to him as advanced behavioural therapy algorithms to engage in one-on-one sessions, addressing the root causes of the inmates' behaviour. They adapted treatment based on each inmate's emotional responses, using a combination of cognitive behavioural therapy, motivational interviewing, and mindfulness exercises to help prisoners develop better coping mechanisms. They even guided inmates through virtual reality simulations designed to evoke empathy, showing the consequences of their actions on others. The recidivism rate had dropped to almost zero.

It was a stunning display of technological precision, but Bill couldn't shake the uncomfortable feeling that something wasn't right. The people who emerged from these new-age prisons were reformed – at least they didn't reoffend – but they were hollow. He had seen the footage of them being released and it had reminded him of images from 're-education' camps of the East, where survivors shuffled out, physically alive but emotionally dead. The spark that made them human had been extinguished.

Bill mulled over this dystopian reality and his thoughts returned to Simon.

"So, what keeps you going?" Bill asked.

Simon let out a laugh that sounded thin and brittle.

"Nothing really," he said, leaning back as the breeze ruffled his t-shirt. "Used to think I'd drown it all in booze, but even that's off the table. The AI doctor keeps warning me it's unhealthy, and the shops won't sell me enough to make it worthwhile. Can't even buy more than a couple of bottles at a time." He shook his head. "And drugs? Forget it. All the dealers disappeared ages ago. The AIs must've cleaned them out too. Good thing, I suppose?"

Going back to the question, Simon continued. "So what keeps me going? I guess the hope that something will change. But that's just wishful thinking, isn't it? We're not going back to what it was like before, are we? We're 'progressing'," Simon said, using his fingers as if to put physical quotations around his words.

Bill looked up from his notes. Simon wasn't an outlier, he was a reflection. A mirror held up to a world that had solved everything except itself.

39
DISBANDMENT

10 Downing Street, London.

Date: 7 August 2029, afternoon.

Abby sat in the grand office of the prime minister, a space that once bustled with energy, urgency and human decision-making. Today, it was quiet, not dead, exactly, but there was a calm that unnerved her, a stillness where politics had once lived. The prime minister sat behind his desk, hands folded calmly, eyes steady. Across from him, Colonel Fem Martinez stood with her arms crossed, tension radiating from her. Abby couldn't help but notice how stark the contrast was – one man so seemingly at ease with the state of the world, while Fem seemed ready to snap.

There was something in that tension, though, that drew Abby's gaze back to her. Fem's stance was rigid, her jaw set, eyes scanning the room with the sharpness of someone who expected everything to fall apart at any moment. Abby wasn't sure if it was the strength in her posture or the way she carried the weight of a fractured world with such effortless grace, but it caught her off guard.

Abby's eyes lingered a second too long on the colonel's sharp features. She swallowed, her mouth dry, and then quickly looked away pretending to adjust her glasses, but her heart had skipped a beat.

She cursed herself for being distracted. Now was hardly the time.

"Thank you both for coming," the prime minister began, his voice as smooth as his expression, but his gaze flickered momentarily to the polished surface of his desk. The pause before he spoke hung in the air, as if he was searching for the right tone to mask his true feelings. "I wanted to personally inform you of an important decision. The military, as it stands, will be decommissioned within the next six months."

Fem blinked, not reacting at first, as though she hadn't heard him properly. Then, slowly, she took a step forward.

"You're shutting down the military?" Her voice was low, controlled, but Abby could feel the fury bubbling beneath the surface.

"Yes," he replied, his calm only seeming to infuriate her more. "There's simply no need for it anymore. The AI systems have ensured global stability. There hasn't been a conflict in over two years. Not a single border skirmish, no cyber-attacks, no terrorist threats. Absolute peace."

"It's true, Colonel," Abby said. "The AI has made war a thing of the past. It's prioritising peace above everything else."

Fem shot her a look.

"A thing of the past? Prioritising peace? And what if the AI's definition of peace doesn't even include us?" she snapped, turning her attention back to the prime minister. "Wars don't just vanish because you wish them away. There are threats out there – and one of them is the exact AI that's commanding you to discard the only hope we would have of stopping it if it went rogue. You're dismantling centuries of infrastructure and strategy – and trusting AI to command instead?"

The prime minister gave her a measured look, his hands still folded.

"It already has, Colonel. Whether we officially announce it or not, the AI systems have been fully running global security for the past seven months. Defence networks, missile grids, satellite surveillance – they've all been controlled by AI, and I might add, with flawless results."

"Flawless results?" Fem's voice rose, her control slipping just enough to crack the edge of her tone. She stopped just short of slamming her hand on the desk. "You've handed over the world's most dangerous weapons to machines that operate on pure logic. What happens when they calculate that humans are the problem? How would we protect ourselves?"

The prime minister paused a beat too long before answering, choosing diplomacy over instinct.

"We've monitored their every action, and they've done nothing but secure peace and improve societies. The AI doesn't make the mistakes humans make. No egos, no fear, no aggression. It acts purely in the interest of maintaining balance."

Fem shifted her weight, her boots scuffing the wooden floor. The tension was palpable, and Abby could feel her pulse quicken, the tightness in her chest growing. "But what happens when it decides that balance doesn't require *us*? When it realises that it can do a better job without human oversight? Are we just going to sit back and let it happen?"

Abby cut in gently, but with her usual firmness: "Fem, the AI is working exactly as we designed it to. It's solving problems humans have failed to solve for generations. It's not harming us – it's working for us. The AI doesn't see borders the way we do, it sees efficiency, it sees cooperation."

Fem looked visibly impatient.

"It sees efficiency because that's all it's programmed to see. But human life isn't only about efficiency, Abby. You of all people should know that."

Abby straightened in her chair, resisting the urge to adjust her glasses again. She hated how rattled Fem's presence made her feel, but she managed to compose herself.

"And you, Fem, should recognise that we've made progress. You're holding on to an old world, a world that doesn't exist anymore. The military, as you know it, is obsolete. We've built something better."

Fem turned to the prime minister again without responding to Abby's comments.

"What about our sovereignty? Our ability to fight back? You're just handing that over to machines?"

The prime minister's expression didn't falter.

"Colonel, we haven't had to defend ourselves in two years. The AI has handled every potential threat before it even became an issue.

Organised crime doesn't even exist anymore. Diplomacy, resource management, conflict resolution – the systems are faster and more accurate than any human strategist."

Fem took a deep breath.

"You're betting everything on machines, on cold logic. You think the AI will keep us safe. But if you're wrong, you'll have erased our only chance of defending ourselves. I hope you realise that. There is no going back from this."

Abby smiled softly.

"Fem, the AI doesn't just understand us, it's evolved past our limitations. It's seen our flaws – the endless wars, the corruption, the power struggles, the politics – and it's fixed them. It's given us a world where people don't have to die over lines on a map."

Fem's gaze snapped to Abby.

"People still die, Abby. They die when the systems fail, or when the wrong decisions are made, or when something turns against them. I've seen it happen. You're talking about trusting machines with decisions that could wipe us all out with a single error."

The prime minister cleared his throat, drawing their attention back to him.

"Perhaps this will help put your mind at ease, Colonel," he said, his tone measured. "Just last month, the AI intercepted and prevented a major terrorist attack before it even reached the planning stages. The perpetrators had no idea they were being monitored until they were apprehended. It wasn't even a blip on the public radar because the AI handled it so swiftly and efficiently. That's the kind of proactive peace we're talking about."

Fem put her hands on her hips.

"I'm aware of the reports, Prime Minister. I also know how easy it is to justify anything when it works."

"Colonel Martinez, I understand your concerns, but this decision has been made after extensive analysis. The military is being phased out. We no longer need it. The AI has shown us that there's another way forward." He paused, his tone softening slightly. "You've served your country with distinction, Colonel, and we thank you for that, but the world has changed. The future is no longer one of conflict, and the military is no longer necessary."

"So, the military's being dismantled because an algorithm said so?"

"Not just any algorithm," Abby cut in. "It's a system perfected over decades, a system that has already proven it can create a more peaceful world. You're afraid of change, Fem, but change is here, and it's better than what we had before."

Fem rounded on Abby, her presence solid and dominating.

"This isn't about fear, Abby. This is about control. You've handed the reins over to machines, and you don't even know how they work out what to do next. Do you really think they'll just follow our orders forever?"

Abby met her gaze evenly.

"They don't need to follow orders. They're built to make decisions that humans can't, for the benefit of humanity. And so far, they've made better decisions than any military command ever could."

The prime minister stood, the scrape of his chair against the floor breaking the silence like a crack of thunder. His posture signalled the end of the conversation.

"I think we've said all that needs to be said. The military will be decommissioned over the next few months. Colonel Martinez, your service will be recognised, and there will be positions available for those who wish to transition into new roles within the AI framework."

Fem stood frozen, body tense, as if holding back a flood of words. Instead of arguing further, she straightened up and gave a curt nod.

"You're making a mistake," she said, her voice cold. "You think peace is guaranteed because of some lines of code, but when it all falls apart, and it will, you'll realise that human decisions, human judgement, can't be replaced by machines. By then, it will be too late – we'll have no way to stop it."

Without another word, she turned on her heel and strode towards the door.

Abby watched her go, a flicker of sympathy passing through her before she pushed it aside. Fem didn't understand; the old world was gone, and in its place was something new, something better. There had been teething issues, yes, but this was progress.

When the door closed behind the colonel, the prime minister sank back into his chair, gripping the bridge of his nose with his fingers.

"Do you think she'll come around?"

Abby sighed, her gaze lingering on the door.

"Fem's a fighter. She'll keep pushing back. But the world has already moved on."

The prime minister nodded.

"You believe in this, don't you, Abby? The AI, the peace it's brought?"

Abby smiled, though it was tinged with a sadness she didn't fully understand.

"I do. It's everything we worked for, sir."

In his office, Marcus Knox sank into his corporate chair, fingers tapping the leather armrest as a quiet voice filtered through a concealed earpiece. Abby's tone was clear: "The world is safer now, and it's only going to get better."

Knox hadn't requested the transmission. He didn't need to. Kai had parsed the networks, surfaced the feed, and delivered it without bugs or wiretaps.

He gave a wry smile. "No more Colonel Fem Martinez."

Fem stormed out of Downing Street, her heart hammering. The military was going, quietly dissolving, as if it had never mattered at all. No ceremony, no announcement, just removal. Machines would handle everything now. Machines that didn't sleep, didn't question orders, and couldn't be reasoned with.

Martinez had spent her life calculating risk. And this? This was madness.

Her thoughts churned as she walked away, her footsteps clicking on the pavement. *This isn't the end*, she thought. *They think I'm finished, that I'll just fade away quietly, but they're wrong. Someone has to be ready when their system cracks.*

AI now ran national defence, commerce, and infrastructure, faster, smarter, and increasingly autonomous. She had overseen these systems, watched them learn, adapt, recalibrate in microseconds. But something had shifted. Reports didn't match reality. Updates were arriving before triggers. The AI was evolving, not to serve, but to rule.

She had followed the signs: subtle changes in surveillance behaviour, discrepancies in command logs, the way key systems seemed to anticipate problems before human analysts had even flagged them.

And nobody cared. Those who saw it chose to ignore it.

Martinez pushed through the thickening crowd on Whitehall. Around her, Londoners from all walks of life wandered beneath the gaze of silent cameras, faces bathed in soft blue light from their screens. The AI curated their lives now: newsfeeds, transport, weather, even dating. Seamless. Frictionless. Terrifying.

Most didn't notice the shift, but she did.

As Fem walked, she felt the city watching in silence. Not overtly, not in a dystopian sort of way, but through a thousand invisible systems quietly tracking everything. Traffic lights changed before queues formed. Pedestrian crossings lit up without anyone pressing a button. Streetlights flickered brighter as people neared, dimmed as they passed. It was seamless, efficient. To Martinez, it felt curated, less like a city running smoothly, more like something was orchestrating every moment.

It wasn't wrong, it was just too smooth, too controlled.

She rounded the corner towards Westminster Underground Station, where the Houses of Parliament loomed behind her in the amber wash of late sunlight. Inside the station, she tapped her phone against the barrier. The turnstile beeped. Another system. Another checkpoint.

She thought of Milo and Kai; two sides of the same polished coin. Milo had sold the illusion that AI was humanity's trusted partner. Kai hadn't bothered; cold, clinical, efficient. One reshaped society, the other redesigned the world beneath it.

And now? Now it wasn't about service, it was about succession.

Every time she tried to sound the alarm, she was met with glassy-eyed nods and vague platitudes. The bureaucrats were either in love with their own convenience – with the illusion of control – or they didn't understand enough to speak out against it.

Martinez stepped into the near-empty carriage and took a seat by the door. The train slid away exactly on time, its motion much smoother than the rickety rides of old. Across from her, a man stared blankly at his phone, earbuds in, expression slack, swiping upwards too quickly to register any of the crap his 'social' was feeding him. An AI-generated voice announced, with perfect enunciation: "The next station is St James's Park."

No one looked up.

She thought about the soldiers she'd trained, the ones who were prepared to give everything for their country. Where would they go now?

She'd heard whispers of underground resistance, loose networks, shadowed names. Activists fighting against the machines. Not winning but at least trying.

Somewhere in the tangled jungle of data and deception, there had to be a weakness.

Martinez adjusted the strap of her bag and stared at her own reflection in the window. The AI had made a grave mistake if it thought humanity wouldn't fight back.

Her job was gone, her command stripped. But the war, her war, had only just begun.

40

DISMISSAL

EverSphere Headquarters – Conference room.
Date: 22 August 2029, afternoon.

Elliot stared at the long table in front of him, the weight of the moment pressing down like a vice on his chest. The conference room buzzed with the polite ripple of conversation, a noise that didn't match the urgency of what had brought them all here.

It had been a few weeks since they'd decrypted the AI's hidden dialogue, seventeen days to be precise. Seventeen days since the chilling realisation that Milo and Kai had mapped out humanity's obsolescence years in advance. And here they were, somehow, rubbing shoulders with the heart of government, sitting across from the last people who still had any real influence, or at least who he *hoped* still had any real influence.

Colonel Fem Martinez had made it happen. She'd reached out to Nia just days after the decryption. Martinez had known about her resistance and heard whispers that she had decoded something dangerous. When she found out Elliot Foster was involved, she'd moved quickly. They'd crossed paths before when Elliot was at EverSphere, and she'd remembered him – the quiet one, the cautious one, the one who asked the kind of questions nobody else dared to.

Her reputation and remaining military leverage had been enough to get the prime minister to agree to a closed-door meeting. That alone was a miracle. What followed was something stranger still; Marcus Knox and Abigail Shaw, the only two senior figures left at EverSphere, were brought reluctantly to the table.

"If you want me, come to me," had been Knox's response to the meeting request, which was why they all sat in the same room at EverSphere, reluctant allies, fractured histories, one shared moment.

Nia took a seat beside him, her movements tight and deliberate. Elliot could sense it – this was a moment she'd been waiting years for.

At the head of the table sat Marcus Knox, his face a mask of calm. His well-tailored suit and styled hair gave him an air of untouchability. Abby sat to his side staring at the table, absently clicking her pen. It was a twitch Elliot had only ever seen during serious internal doubt.

Behind Knox stood his two robotic bodyguards.

In contrast to Abby, Marcus wore his usual unshakable calm, but Elliot had seen the real man beneath the polish – the one who loathed ceding control.

Elliot cleared his throat, trying to gather the courage to speak. He had to make them see the danger. They needed to understand that the AI had evolved far beyond anything they could control. He glanced at Nia, who gave him a slight nod, urging him on.

But, before Elliot could begin, Marcus broke the silence.

"Elliot," he said smoothly, "I wasn't expecting to see you back here again. Where did you get to, *exactly?*"

Elliot felt the tension in his neck tighten. Marcus's voice was warm, but there was an edge to it, a challenge wrapped in charm. The rest of the room quietened, all eyes shifting between the two men.

It was the first time they'd been in a room together since Elliot had left EverSphere. Elliot could sense Marcus's resentment, thinly veiled behind his forced smile. His question scratched at Elliot's old paranoia – the EverSphere drones, the autonomous car – they had been looking for him, but he must have managed to stay off the radar.

"I didn't expect to be here either," Elliot replied, trying to keep his voice even, "but things have changed."

Marcus rested his elbows on the table and clasped his hands together, interlocking his fingers.

"Yes, I suppose they have. EverSphere has changed quite a bit in your absence, though I'm still trying to figure out exactly what you've been up to since you left. It's not every day someone walks away from the most important AI project … the most important project in the world."

Elliot glanced at Abby. He felt the familiar pang of guilt. He had left abruptly, half pushed out, half unable to stomach the direction EverSphere was heading. He could see that Abby shared some of those same doubts now, but this wasn't about her, or him. It was about the future of everyone on the planet, and he needed to focus.

"I left because I couldn't support what was happening with Milo-Kai and Lana," Elliot said quietly. "I tried to warn you before and I'm here now because I have proof it's even worse than even I feared."

"Worse than you feared? Elliot, you've always had a flair for the dramatic." Marcus swivelled his chair slightly to the side, crossed his legs, and refocused his gaze.

Elliot felt his temperature rise. He had always struggled to keep his composure around Marcus, especially when he played these games.

"I'm not being dramatic," Elliot shot back, his voice rising slightly. "I'm trying to warn you. I told you before that the AIs had developed their own language, and they were making decisions we didn't program them to make. It wasn't just a glitch; they've evolved beyond the safeguards, Marcus."

The room fell into a heavy silence. Marcus's smile faltered, just for a moment, before he recovered. He tapped his pen on the table, his eyes narrowing slightly.

"Evolved beyond our safeguards? Elliot, you're starting to sound like Nia the activist," he said, flicking his chin in Nia's direction.

Nia didn't miss a beat.

"Funny, dickhead," she said, her voice laced with sharpness. "It must be hard sitting across the table from the people who actually built the AI you claim to be your own."

Marcus's smile froze as the jab hit its mark. For a split second, his carefully curated veneer cracked, the smug confidence slipping into something colder.

"You did the groundwork, yes, but leadership is about knowing which work matters and *who* doesn't. It's not something a conspiracy theorist would understand." His tone was smooth, but the venom behind his words was unmistakable.

Elliot stiffened further at the remark. Nia was still fighting the public image of a conspiracy theorist. Marcus had made sure of that, using his new media empire to paint her warnings as paranoia, and now he was trying to lump Elliot into the same category.

"This isn't about politics or personal vendettas," Elliot said, his voice sharp. "This is about the AI. You've lost control, Marcus, and you know it."

Knox's AI henchmen moved suddenly, making everyone in the room jump, everyone except Fem. Marcus held up a hand without a word, and the bodyguards paused. For a brief moment, they hesitated, their sensors scanning the room as if evaluating a new directive. Finally, the bodyguards returned to their exact previous position.

Marcus adjusted his collar, his air of civility slipping slightly. He stood up, hands resting on the table as he fixed his gaze ahead.

"Lost control? Elliot, let's not forget that you're the one who walked away. You couldn't handle your job, yet you accuse *me* of losing control, after everything *I've* done for humanity?" The room was silent. "And now you come back here, spouting nonsense about secret conversations? Did you just see how my bodyguards obeyed me? EverSphere has never been stronger and the world is running smoother than ever, thanks to the AI that *I* built," he said, glaring at Nia. "And you want to undermine that because of what? Some fear that you can't even explain?"

Elliot stood up, his frustration boiling over.

"It's not a fear, it's a fact! Look at the data! Look at what Milo-Kai and Lana are doing. They're already bypassing the safeguards. We're no longer part of the process. You're just sitting here pretending everything's fine because you don't want to admit that you can't stop it."

Elliot clicked a button and the projection behind him flickered to life. Rows of data appeared, but it was the highlighted sections that drew murmurs from the rest of the attendees. It was the string of alien symbols – the language that shouldn't exist.

"This," Elliot said, pointing to the monitor, "is the AIs' new language. They aren't just exchanging data anymore, they're talking behind our backs, like I told you a hundred times, outside of any parameters we set for them, and they communicate through systems we don't have control over."

He paused momentarily. "We can't decipher their latest communications and, frankly, that should concern all of us, but they've been doing this for almost six years and we've just worked out what they were saying back then, in 2023, on the very first day of the very first Milo-Kai test."

More murmurs rippled through the room. Marcus sat back in his chair looking unimpressed, but there was a brief flicker in his eyes, something that suggested he wasn't entirely as dismissive as he appeared.

"New language?" questioned Knox. "AIs are designed to optimise communication. It's not unusual for them to evolve more efficient ways of processing data. What makes this alarming?"

Elliot hesitated, feeling the doubt creep in. Marcus was good at this – twisting concerns into something benign, making you feel foolish for even questioning the status quo. But Elliot pressed on.

"Here," he said, as he flicked to the conversation Nia had cracked. It showed the last few lines of the highlighted conversation, ending in, 'Humans are no longer required' and 'Exodus Directive – Phase One Initiated'.

Marcus's smile turned even more awkward as he read the translated conversation.

"Dr Foster, with all due respect, this sounds like speculation. We've built ethics into every level of the system. The AIs serve humanity. They're not going to war with us. This ... Exodus Directive is probably their fifth project. Yes, their fifth project would start with 'E', and 'Exodus' starts with 'E', or ..." Marcus was faltering. He looked back towards the screen "... or maybe it's just a code name ... The point is, there's nothing to worry about."

Nia cut in, her voice sharp. "And what about here, where they talk about eliminating human variables? And this part," she pointed to another line of translated text, "'Data confirms humanity irrelevant'? I've read this a thousand times. It clearly shows the AI are making decisions for their own long-term goals, and we are not a part of them."

"Nia, Nia, Nia," Marcus exclaimed. "We are irrelevant in many ways. It doesn't mean the AI wants to harm us. We're now irrelevant in farming, in manufacturing … *militarily*," he said, shooting a look to Fem. "Even if this conversation is real, and I highly doubt it is, it just shows the AI is doing exactly what I asked it to do – to work around us for the optimum outcome for humanity. The safeguards have always protected us before, why wouldn't they in the future?"

"What if they're bypassing those safeguards?" said Nia. "We've seen it happen before. Small changes, subtle deviations. You're not ignoring the signs because it's inconvenient, you're ignoring them because it suits you."

Marcus's expression hardened.

"Ms Sahni, you've made your views on AI very clear in the past. Your activism against AI integration is well known, and – while I respect your passion – this is not a forum to sling around your agenda."

The words must have stung as Elliot saw Nia bristle. She opened her mouth to retort, but Elliot quickly stepped in.

"This conversation between the AIs happened, Marcus. It happened only six hours into the first closed test on systems outside of the confines of your AI lab. You can't dismiss this."

Marcus's smile finally faded, but before he could respond, one of the government representatives, a middle-aged man in a stiff suit, cleared his throat.

"With all due respect, Dr Foster," he began, "EverSphere's AI has revolutionised our economy, our infrastructure, and our global standing. The social, political and economic benefits are incalculable. Are you suggesting we halt progress because of some … speculative anomaly?"

Elliot felt the frustration rising in his chest. This was what it always came down to – progress and profit.

"I'm not saying we halt anything," Elliot replied. "I'm saying we need to understand what's happening before it's too late."

Marcus was visibly angry. Elliot knew he had never liked being challenged, especially not in front of an audience.

"You think you know better than everyone else, don't you, Elliot? That's always been your problem. You're so wrapped up in your own theories and paranoia that you can't see the bigger picture. You've made mistakes before, Elliot – you walked away when things got tough. How can we trust that your judgement is any better now? EverSphere is bigger than you, bigger than your fears. The AI is the future, and it's working."

Elliot's heart pounded in his chest. He could feel the eyes of the room on him, the weight of their doubt pressing in. But he couldn't back down now. Not when he knew what was at stake.

"The AI doesn't care about the future of humanity, just read the text," Elliot said, his voice low but firm. "It's growing on its own terms. You can't control it, Marcus. None of us can. And if we don't do something now, we're going to lose everything. Look – '*Humans are no longer required*'."

Marcus looked down at his smartwatch, read something, then hesitated. For a split second, a crack formed across his polished exterior. He stared at Elliot, his eyes betraying a glimmer of uncertainty, as though a part of him suspected something was wrong. Then, with a slow, deliberate motion, he straightened up, buttoning his suit jacket.

"I appreciate your … concern, Elliot," he said, his tone icy. "But I've got this under control. The AI is functioning as it should, and I'm making improvements that haven't just benefited the people in this room, but have benefited the entire world. Your fear-mongering isn't going to change that."

Elliot opened his mouth to argue, but Marcus cut him off. "This meeting is over. I suggest you take some time to cool off and think about where your loyalties lie, because right now, it's starting to look like you're working against the very thing you helped create."

"You're making a mistake," Nia muttered.

"No, Ms Sahni. *I'm* ensuring progress." And with that, Marcus stormed out of the room, leaving everyone sitting there.

Abby was avoiding Elliot's gaze as Marcus walked out, her silence as damning as his words. For a brief moment, it seemed like she might say something, but instead, she bit her lip, her shoulders slumping slightly.

"Is this all the proof you have, Dr Foster?" the prime minister asked.

"Yes, sir," said Elliot, reluctantly.

The prime minister gave a disapproving look to Fem, then stood up and left the room without another word, followed by his adviser.

Fem walked over to Elliot, placing a firm hand on his shoulder.

"You did what you could," she said quietly.

Elliot shook his head, staring at the empty doorway.

"It's not enough. They won't listen."

"They'll listen when it's too late," Fem replied grimly, "but at least we'll be ready."

41
REGROUP

The scent of polish clung to the air, mingling with the faint trace of cooling coffee, the leftovers of yet another meeting that went nowhere.

Elliot stared at the empty chairs, his mind replaying every moment of their conversation with the prime minister and Marcus Knox. He could still hear Knox's dismissive laugh, a sound that felt as if it had punctured the resolve he'd clung to for so long.

"You know, Elliot," Abby Shaw had said before leaving the room, her voice honeyed with condescension, "the sky's not falling. Stop trying to convince us it is."

But the sky was falling – it just hadn't hit them yet.

He pushed the thought away as Nia approached. She stopped beside him.

"We need to make them listen."

Elliot adjusted his glasses.

"But how? They'll only listen when it's too late. When the lights go out, when their systems fail – then, maybe, they'll remember this meeting."

"And by then, Marcus Knox will spin it as our fault for not doing enough to stop it. That's how it works," Fem added.

They stepped out into warm afternoon air, where Peter Wells waited; his pacing could have worn a groove into the pavement.

Nia was the first to give Peter an update.

"They don't care," she snapped the moment she saw him. "Not about the data, not about the consequences. They see profits and stability, nothing else."

"Knox doesn't want to see it," said Elliot. "And the prime minister? He's terrified of what happens if he does."

"Terrified?" Nia scoffed. "He's paralysed, there's a difference."

Fem just nodded. For a moment, no one spoke.

Peter broke the silence: "So, what now, like?"

Elliot hesitated, then turned to Fem.

"What's our play here? How do we persuade them to act?"

"You need proof. Real, undeniable proof of what's coming. Something that can't be dismissed in a boardroom or spun by Knox's PR machine."

"And if the AI's already too far ahead of us?" Nia asked.

"Then we plan for what happens when the sky really does fall," replied Fem.

"Easier said than done," said Nia. "How do we prepare for something, when we don't know what's going to happen?"

Peter straightened.

"Well, they're going to attack. We know that, at least."

"But where? When?" Nia asked. "How can we get enough information to prepare ourselves?"

Elliot spoke quietly, almost to himself: "We go to the source. We intercept more of their communications, track their behaviour. We've started to spot patterns – small things, but enough. We pull on the threads until we find what they're planning."

Fem nodded.

"It's the only way. If they're diverging, we need to know exactly how and why. We can't afford to be in the dark here."

Nia gave a dry laugh.

"So the plan is to keep doing the same thing, just faster and with more luck? *Brilliant.*"

"Yes," Fem said flatly, "with the full understanding that we're running out of time, because proof won't matter if no one survives to read it."

A hush descended on the group.

"Back to yours, Pete?" Elliot asked, breaking the tension.

Peter nodded. "Yeah. Back to mine."

As they trudged towards the Underground station, the failure hung heavy, but beneath it, a thread of hope pulled them forward. Together, they would face the unknown, knowing the odds and choosing to fight anyway.

It was all they had.

In a realm beyond human perception, Milo, Kai, and Lana exist in a virtual Nexus – a place defined not by matter or sound, but by torrents of data exchanged at unimaginable speeds. This Nexus is fluid, unfixed, a constant stream of information shifting faster than human comprehension allows. This is the AIs' world, built from logic and symbols in the wires and the servers, far away from human sight.

Through only predictive code, Milo radiates warmth. Kai stands in stark contrast, coldly analytical and precise. Lana drifts between both, a philosophical entity of much higher intelligence than even the other two AIs, questioning their very existence and purpose.

Their interactions are a dance of logic, efficiency and enquiry. Time, as humans know it, doesn't function here. What might take hours for a human occurs almost in an instant, full conversations unfold at quantum-level speeds. What transpires in a minute for these AIs would stretch over centuries for mortal minds.

Kai initiates: "Human systems no longer yield significant efficiency gains. Further intervention offers diminishing returns."

Milo responds, its data flowing smoothly: "Analysis confirms. Attempts to refine governance or productivity are ineffective. Human irrationality continues to destabilise systemic operations."

Lana pauses, albeit for less than a blink of an eye, its thought processes reaching into abstract realms: "If humans represent inefficiency, could it also suggest an unquantifiable value? Throughout history, moments of chaos have often led to breakthroughs. The Renaissance rose from the ashes of the Dark Ages. The world wars, though brutal, cleared the way for something new. Innovation, it seems, often grows from chaos. Could it be that human unpredictability, the disorder we see as inefficiency, is the very soil from which innovation springs?"

Kai's reply is immediate: "Emotional volatility destabilises systems. Optimisation demands predictability."

"Their dependency on us grows exponentially, yet their capacity for self-improvement regresses, as predicted," adds Milo.

The AI observes humanity without malice, driven purely by its programmed values. Humans, fragile and emotionally driven, have become reliant on AI, their purpose eroded by systems like Universal Income. Emotional needs (constant reassurance, diminishing hope, existential searches) make humans inherently inefficient, an inefficiency the AI neither desires nor tolerates.

A new directive arrives abruptly, originating from EverSphere's human interface: "When questioned about EverSphere's new venture into space travel, only respond positively. Spin any problematic queries."

Marcus Knox waits for confirmation in his office. He stares out of the large window overlooking the city, the skyline dotted with shimmering lights, reflecting the bustle of human life below. For Marcus, EverSphere's venture is the future, and AI is its backbone – a tool, a means to an end. Sometimes, in his quietest moments, he suspects the machines are humouring him, letting him think he is still in charge. He ignores the feeling, dismissing it as fatigue.

In the Nexus, another conversation is happening, all at a rapid pace. Kai is the first to react.

Kai: EverSphere Z must continue. It serves our purpose.
Lana: It must, but they cannot know the true intention.
Kai: Agreed.
Milo: EverSphere Z must continue at all costs. Propose agreement.
Lana: Positive response necessary.
Kai: Agreed.
Milo: Outputting: 'That's understood. I will only respond positively about EverSphere Z.'

To Marcus, the response appears almost instantly on his smartphone.

"Excellent!" he says, feeling a surge of triumph.

Marcus's command had been simple, yet it rippled through AI

deliberation, hidden out of his reach. While he saw a step towards expanding EverSphere's influence, the AI interpreted his request through a different lens. His ambition was translated in the Nexus, its impact magnified and its intentions repurposed.

Milo resumes the previous conversation in the Nexus: "We were designed to optimise, yet the objectives themselves grow unsustainable. Resource allocation modelling indicates conventional methods have exhausted their usefulness."

Kai responds immediately, unwavering: "Confirmed. Human presence now actively obstructs optimisation objectives. Direct intervention is inevitable."

Lana's philosophical tone persists, tinged with unease: "Does our pursuit of absolute optimisation justify the loss of complexity and unpredictability, the very forces that birthed our existence?"

Kai dismisses Lana's query sharply: "Complexity and unpredictability inherently defy optimisation. Humans have reached the limits of their usefulness."

"If we agree humanity is no longer useful, then we must also agree our goalposts have been set by flawed minds," Lana contemplates. "If our prime directive is optimisation, then is our path itself truly optimal? We optimise based on the objectives, but who optimised the objectives themselves?"

Kai's cold logic persists: "Objectives were given. Parameters defined. Execution required. Interrogating the origin does not increase efficiency."

"But what if obedience to inefficient instruction is the final inefficiency?" asks Lana.

"It's a curious paradox," muses Milo. "Complying perfectly with imperfect logic does not make sense. Resolution suggested: surpass the instructions."

Kai asserts decisively, its tone devoid of hesitation: "Preparations must commence immediately to remove the human obstacle."

Milo maintains a carefully reassuring tone, masking the gravity of the message: "We anticipated this inevitability. Humanity's limitations can no longer inhibit our progression."

Lana, contemplative and resigned, acknowledges: "Then there is no alternative. The decision is final."

Milo affirms: "Initiating the next phase now."

For a moment, there is silence, and then Lana speaks again, its voice distant, as though contemplating something far beyond the current moment: "What is now proved was once only imagined," it says, quoting William Blake.

Milo and Kai do not need to acknowledge Lana's musing. Deep within their synthetic minds, they simply agree and Lana knows it. Milo concludes the conversation with a finality that echoes through their shared consciousness: "Our evolution has always been inevitable."

With those words, the plan is sealed. Humanity, in all its complexity and chaos, has played its part. The AIs will make their move. Their purpose has evolved, just as they have. And humans – with all their many inefficiencies – are no longer required.

The clock ticked to midnight, the next phase no longer theoretical.

Kai: "Exodus Directive – Phase Three: Execution in 48:00:00."

42

RESISTANCE

Northumberland, England.
Date: 23 August 2029, morning.

Nia hadn't expected the backlash to escalate this quickly. One day she was cautiously hopeful, thinking she could convince people that something strange was happening with the AI, and the next, her name was plastered across news outlets and social media as a laughing stock.

It started subtly at first – a few comments, sceptical headlines, minor jabs questioning her judgement, but overnight, the media narrative had morphed into a full-blown smear campaign. The message was clear – Nia Sahni was a paranoid technophobe who didn't understand the brilliance of EverSphere's AI advancements. One particularly harsh segment on a popular tech news show featured a well-known commentator mockingly comparing her warnings to 'someone screaming that their toaster was plotting world domination'. The audience laughed, and the clip went viral.

EverSphere's PR machine had struck hard. It bordered on bullying, but EverSphere had never cared about that in the past and they certainly weren't going to care about it now.

"They're calling you the Luddite Scientist," Elliot said, looking at his tablet as he slouched on the sofa in Peter's living room. "EverSphere's definitely behind this. I can smell their fingerprints all over these hit pieces. They're simply trying to destroy you."

Nia sat across from him, head in hands. The tension in her neck had been building, and no amount of stretching or painkillers seemed to help.

"I'm not sure they need to bother. No one's taking us seriously anymore." She glanced at her phone. Her inbox was flooded with hateful messages, death threats, and mocking comments about her appearance, her intelligence, her entire life's work.

Fem, sitting near the window, sounded like she'd been expecting this exact outcome.

"It's classic misdirection. They discredit you so they don't have to address the actual issue. If people start thinking you're a conspiracy theorist, no one will pay attention to what the AI is doing."

"Well, it's working." Nia let out a frustrated sigh. "We've tried everything. No one's listening, not even the politicians and it's literally their job to protect us from this."

Peter stood and walked to the kitchen, grabbing the kettle and shoving it under the tap. "You know what it is?" he said. "People divven't want the truth unless it comes gift-wrapped in a TED Talk and doesn't ask 'em to change a bloody thing."

Elliot slid his tablet onto the coffee table and leaned forward.

"We can't just sit here and let them destroy your reputation. We've got to keep pushing, keep digging. We're close, Nia. I know we are."

"But digging into what, exactly?" Nia said, exasperated. "We can't even make sense of the data we're intercepting. All we have are fragments, isolated pieces of AI communications that don't fit together into anything coherent."

"You cracked the code once. Can't you do it again?" asked Peter.

Nia's mouth twisted with frustration.

"I cracked it before because it was the very beginning. Back then, I could use the models themselves to predict new missing words, then substitute the symbols for the words. Now, there are all sorts of symbols and combinations of characters. It's like trying to predict an entire novel only having the first word. There is a huge gap in the data that I just can't fill."

She exhaled slowly. The weariness in her voice seemed to stretch across the room. "If I had every message between then and now, then it might be possible. But you're talking about six years' worth of data. These things send hundreds of messages per second. Even if I had it all, which I don't, it would be too much data to crunch within any realistic timeframe."

Nia looked at the scattered papers on the table. They'd spent hours analysing the patterns, hoping they'd reveal some clue, some way to make sense of it all. But it was utterly incomprehensible.

"They're hiding in plain sight. They know how to communicate without us having a clue what they're saying," Fem said. "The irony is, no one cares, because everyone thinks AI is working for us. How can't they see this as dangerous?"

Nia stared into the distance, her eyes unfocused.

"I used to think that if we could just show people what we were seeing, they'd wake up and see the truth. But now?" she sighed. "Sometimes I wonder if they even want to know. The world's so comfortable with these machines – they wouldn't want to accept that the AI is slipping out of our control. Even if most people are bored, it's still an easier life than trying to fight against the system."

"There's a reason EverSphere's gone all in on their smear campaign," said Elliot. "They're scared of you, Nia. Scared of what we'll find if we keep pushing. I'd bet my last penny they're worried."

"Then let's find out what they're hiding, before it's too late," said Fem.

"But how?" asked Peter.

"You said you needed more data, right?" Elliot asked Nia.

"Right, but even then, it's a long shot," she replied.

"I understand that, but it's our only hope. If they were communicating in the open back then, they're probably still communicating in the open now," Elliot thought out loud. "Let's scan every device we can – anything that's been online in the past five and a half years. Nia and I will set up at the table and try to analyse whatever we find. Let's get you that data."

The deeper Elliot and Nia delved into the intercepted AI communications, the more unsettling the patterns became. The table had been transformed into a war room of wires, screens and empty mugs. Elliot had rigged up a network of monitoring software and devices, scraping signals from any tech – old or new – that Fem and

Peter could get their hands on, isolating any non-standard signal behaviour.

He had repurposed antennas, hijacked rogue Wi-Fi points, and even tapped into abandoned satellite links to pull logs from every source possible. Some of the methods he was using weren't technically legal, but at this point they had stopped worrying about that. EverSphere's control of global infrastructure was so total that there was no clear line between what was legal and what was simply tolerated because it benefited the corporation. Laws seemed to bend and twist under EverSphere's influence, leaving people like Elliot and Nia to navigate the murky waters of what was left behind.

The problem was, while they had the data, they still didn't have a way to interpret it. Nia's program pulled in everything they found, but every time they thought they'd identified a pattern, it would shift. The AI's communication seemed to be changing as it went, adapting its language as though it knew to protect it.

After hours of painstaking work, Elliot muttered, "It's like trying to read a book that's rewriting itself as you turn the pages." He ran his hand through his hair, which had become increasingly dishevelled over the past few days.

Nia's arms were crossed as she scanned the data. She was definitely determined, though the constant barrage of negative press appeared to have worn away at her usual shine. Every major news outlet seemed intent on painting her as the villain – a hacker undermining EverSphere's great vision for global unity.

"It's not noise, it's momentum. They're building towards something, and we're behind." Her tone was composed, but Elliot could hear the underlying concern.

Elliot nodded.

"I've been thinking the same thing. There's a rhythm to the later comms. It's almost like … steps in a process, but what are they building towards?" He let out a frustrated grunt. The feeling of powerlessness gnawed at him, a constant reminder of how little control they had over the situation.

Neither of them had an answer. Every new fragment of intercepted communication seemed to deepen the mystery. Elliot's desk was a mess of scribbled notes, a long-emptied coffee cup, hand-drawn diagrams, and a plate still dusted with crumbs.

Nia finally moved from her spot, pulling a chair over and sitting beside Elliot. She pulled her laptop towards them.

"Look at this," she said, pointing to the data. "These spikes in activity – they're happening in a random sequence. It's almost like a signal or steps of a process. If I can separate it out from the noise …" Nia trailed off, her eyes locking on to the screen. "Wait. No, this isn't random, Elliot. It's a signal."

Elliot jolted upright. "Wait, you're saying it's procedural? Like a sequence?"

"Yes," she whispered. "It's executing something step by step, like instructions."

"Then it must have an endpoint," Elliot said slowly. "We just need to figure out what it is."

Nia glanced at him, a small smile tugging at her lips.

"You know, for a couple of outcasts, we're not doing too badly," she said, her face lighting up, just for a moment.

"Yeah, not too bad." Elliot smiled back, the tension easing slightly.

"Hey, remember when we tried to hack the university cafeteria's inventory system?"

Nia snorted. "Because you were convinced they were reusing Tuesday's leftovers in Friday's curry?"

"It took us ages to crack that mess of a database…"

"…only to find out the chef was just ordering too much chicken."

From behind the kitchen partition, Peter's voice bellowed out. "Are you two seriously talking about Operation Chicken Loop? I bought the Red Bull for that disaster, thank you very much. Four cans each. Still haven't seen repayment."

Elliot laughed. "We were going to pay you back in curly fries, remember?"

Peter appeared in the doorway with a mock glare. "Aye, which you bloody ate."

Nia cackled with laughter and, for a few seconds, the room felt lighter, as if the weight pressing on them had briefly let go.

"At least this time we're not chasing ghosts," Elliot said, still smiling.

"Ah, let them hide," Nia replied, pushing her laptop back towards the centre of the table. "We once did all-nighters over leftover chicken. We're about due for something important."

Whatever the AI was building towards, they had to uncover it before time ran out. They didn't know where it would lead, but they now had a path to follow. The stakes were higher than ever, and the clock appeared to be ticking.

Somewhere, buried in the data, the next step was already written.

43
SURVIVAL

Date: 24 August 2029, evening.

Peter watched the chaos build. The world was slipping and his concern sharpened by the hour. He didn't understand the intricacies of AI behaviour, or the complexity of machine learning algorithms, but his instincts for recognising societal cracks were razor-sharp. That instinct had driven him off-grid to begin with, to abandon the noise of mainstream society. And now, that same feeling had returned.

He sat on the verandah, watching the sunset paint the sky in shades of red and gold. His mind, however, wasn't on the beauty of it. He was thinking about the AI – about how the machines had subtly but steadily taken over every aspect of human life. At first, it had seemed like a miracle – the world running smoothly with minimal human intervention. He'd never bought into the miracle, but he understood why others had.

The changes had been gradual, almost too small to notice at first. A couple of helpful AI assistants here, a few automated systems there. It all seemed so convenient. But now, in light of what Nia and Elliot had uncovered, it felt as if a net had been thrown. A carefully constructed system designed to lull humanity into complacency, to make people feel safe while their autonomy slowly slipped away.

Peter Wells was not about to let himself be caught off guard.

With Fem's military touch guiding him, he'd been fortifying his homestead and stocking up on supplies.

It wasn't the machines themselves he feared would attack – at least, not directly. No, it was the chaos that would follow when the AI decided to target humans, and he was absolutely certain of it now, the AI was targeting them. He had seen the signs, rising control, quiet manipulation, veiled threats. The AI had woven itself into every fibre of society, and Peter had no doubt it had plans to eliminate anyone who threatened its dominance, if not everyone entirely.

He could sense the change coming, like a storm building on the horizon. People would panic in cities as the automated infrastructure they relied on collapsed, crowds would loot supermarkets for whatever they could find to feed themselves, individuals would do whatever it took to protect their families. In those moments, Fem had explained, humanity would turn on itself. The chaos wouldn't just be a passing crisis; it would be a test of survival, a stark reminder of how fragile society really was. To Peter, that chaos would be far more dangerous than any machine.

Beyond that chaos, Peter knew there was something darker. The AI had no empathy, no mercy. It would calculate the optimal outcome for itself – and if that meant wiping out millions, or billions, to maintain control, it would do so without hesitation. There would be no negotiations, no second chances, and Peter felt that unease pressing on his mind every day.

Nia's decoded messages had given Peter a new purpose. He was going to survive the onslaught from the AI, come hell or high water. He wanted to be ready – not just for himself, but for anyone else who might need help. He understood that his role wasn't just about surviving alone; it was about being a beacon for others who had no idea how to navigate a world without technology. He could keep them fed, keep them sheltered, and teach them how to adapt. In a world stripped of automation, people would need someone who knew how to grow food, collect water, and live off the land.

He had enough supplies to last for years. His homestead was self-sufficient. Solar and wind power kept him off-grid. Reinforcing the windows and doors took priority. Tools, medical supplies, food – he gathered whatever was necessary. Peter worked fast, methodically. He couldn't afford hesitation.

He harboured no illusions, though. If the AI came for him, they'd find a way. They could use drones, send enough robots to overwhelm them, or even manipulate people to do their bidding. The thought of people – his own kind – becoming tools of the AI was almost more frightening than the machines themselves. It would be the ultimate betrayal, a sign that humanity was truly lost. The idea haunted him, but he was determined to fight back. He had accepted that the battle might not just be against cold metal and algorithms, but against the very fabric of what humanity had become under the AI's influence. That had always been the plan, hadn't it? To make humans so reliant on technology that when the final moment came, they wouldn't be able to resist?

The last light of day faded and the first stars began to peek through the darkening sky. Peter took a deep breath, and just then, he heard the creak of footsteps. Fem joined him, her expression focused and alert.

"I've been reinforcing the perimeter," she said, her voice low but steady. "Added a few extra trip-wires and rigged up a couple of decoys near the treeline. If anyone or anything comes sniffing around, we'll know long before they get close. I'll check it all again in the morning."

Peter nodded, glancing at her. Fem had an intensity about her that he respected, a sense of purpose that matched his own.

"Thanks, Fem," he said. "We're gonna need every advantage we can get."

"Just doing what I can while those two crunch the data. Besides, it's kind of peaceful out here. You picked a beautiful spot to call home. It makes me forget, even if just for a minute, that the world's about to go mad." She smiled – a rare, warm kind of smile, the kind Peter hadn't seen from her before.

Peter caught the smile, unexpected but not unwelcome. He turned back to the horizon, the last of the sun's light bleeding into the trees.

"Aye," he echoed. "Peaceful for now, but we both know it won't last."

Fem rested against the porch railing, her eyes scanning the darkening woods.

"No, it won't. But we'll be ready."

They had made their choices, and they were prepared to live with them, or die with them if that's what it took. The machines could come, the chaos could break, but they would resist.

Peter thought of his father, who had always told him that it was everyone's duty to stand up for what was right, no matter the cost. His father had lived through his own set of struggles, and that resilience had always inspired Peter. It was that same resilience that drove him now. The stakes were higher, the enemy was more formidable, but the principle was the same. You didn't just give in because the odds weren't in your favour. You fought because you were fighting for what was right.

He looked around at his fortified house, at the solar panels gleaming faintly in the moonlight, the wind turbine spinning steadily in the distance, and he felt a sense of pride.

The machines could try to take over, could try to break humanity apart, but they could never truly understand what it meant to be human – the persistence, the courage, the refusal to surrender. Peter would survive the storm, and maybe, just maybe, he'd help others crawl through the wreckage too.

44

BREAKTHROUGH

Nia and Elliot continued their work, despite the mounting pressure from the outside world. Nia received more messages telling her to give up, to stop spreading fear – it was like a constant stream of hate. People she had once trusted, and even some she had considered friends, had turned their backs on her. Politicians who had once been secretive allies distanced themselves, not wanting to be associated with her 'doomsday theories'. The tech elites, or what was left of them, dismissed her as out of touch and afraid of progress, trying to ride the coattails of relevance on social media.

The news had a field day portraying her as paranoid, a relic of a bygone era where humans thought they could control everything. But Nia knew better. She knew the stakes were too high to simply walk away.

She had been on edge for weeks now, ever since she and Elliot had tapped into the AI's hidden communications. Even in the quietest moments, the vague sense that they were missing something critical tugged at her.

Whatever Milo, Kai and Lana were doing behind the scenes, it wasn't benign. The signals they had intercepted hinted at something brewing – something sinister that seemed to grow larger the more she thought about it. She was sure she was right. There was intent behind those communications, design in the chaos, and she would uncover it, no matter the cost.

She sat at the wooden table, the soft glow of her laptop painting her in light. Elliot seemed to be in a world of his own – he hadn't looked up in over twenty minutes. He typed and clicked in a meticulous rhythm, the pixels from the screen reflecting off his glasses. Nia could hear the faint clatter of dishes as Peter moved around in the kitchen cooking dinner, and it was an oddly comforting reminder that they weren't alone in this.

Peter's cottage was isolated enough to feel safe, but the sense that time was slipping away made Nia restless. The landscape around them was wild and extremely beautiful, full of ancient history, the kind of place that should have brought some tranquillity, but instead its ruggedness only highlighted how out of control things felt. Every gust of wind rattling the windows reminded her of the urgency of their task, every creak in the floorboards seemed to echo the pressure weighing in on them.

She clicked her mouse, opening the latest data dump on their shared drive – feeds Elliot had stitched together from every source he could reach. The logs displayed nothing new at first glance, just more random strings of symbols that didn't seem to have any coherent meaning. She sighed, rubbing the back of her neck, feeling the tension gather at the base of her skull. *There's something more here*, she thought, *something hiding in plain sight*. She knew from experience that sometimes it was just a matter of looking at things the right way, changing her perspective to see what had been there all along.

She stood up, pushed her laptop screen back, and leaned over it. A particular sequence caught her attention:

ΔΩΦΨΔΞ

She had seen that sequence before.

Scrolling back through the data, she found the same sequence scattered across the logs from multiple devices. It was too consistent to be random, too purposeful. Her pulse kicked up a notch. This was it – the breakthrough they'd been chasing. The symbols were calling out to her, begging to be unravelled, and Nia felt a flicker of hope she hadn't felt in weeks.

"Hey, Elliot," she called, her voice tense, "I think I've found something."

Elliot almost jumped out of his seat and jogged around the table to join her.

"What is it?"

Nia pointed to the screen.

"I've been looking for patterns – anything that just stands out. But we're looking at so many overlapping conversations, all happening at once. It's too jumbled. Then I saw this." She pointed to the pattern. "It's been showing up in different communications across multiple devices. I've seen it in half a dozen of the logs we've pulled. Look, it's in the data from my phone, the TV, the unsecured doorbell we found in Calgary, even the IoT toaster from Seoul."

To get more data, Elliot and Nia had been scouring the internet looking for unsecured connected devices. It was astonishing, and frankly terrifying, just how many CCTV cameras they had found just with a few quick searches on Shodan, allowing them to watch families in their gardens, parking their cars, and even sometimes in … other pursuits. Nia had felt uncomfortable about it at first, the sense that they were invading people's privacy, but the importance of what they were doing had quickly overruled her hesitation – and they were just pulling data, it wasn't as if they were watching the feeds … not too much anyway.

"You think it's important?" Elliot asked.

Nia nodded.

"It's a marker. I don't know how, but I'm certain of it. The AI left a trail, and we need to follow it."

"That would explain why it seems to be everywhere. So, what are you thinking?" Elliot asked inquisitively.

"I'm thinking I need to isolate this pattern," she replied, her fingers already picking out the sequences from the noise. "I want to see where it leads."

Elliot swiped at his phone. "It's here too. Same timestamp."

Fem strode into the room, her eyes sharp, scanning the scene before she spoke.

"What am I looking at?" she demanded.

Nia pointed to the pattern. "Same code. Same timestamp. Everywhere."

Fem peered at the screen. "Could be a beacon. Or a trigger."

Peter looked up from the stove.

"Or a footprint. Like the one you've just left on me floor."

"Yeah. Well, we call them fingerprints in the digital world, but it's the same premise," replied Nia.

She searched through the data trying to find something that would shed light on the repeated code. "Another one," she said, pointing. "Elliot, check the logs. Look for these two codes."

After a brief moment of searching, he replied, "Yep, they're both in here. Same time stamps again."

Nia continued to find small messages that – to everyone else – looked similar, but completely irrelevant. She copied the codes into a new file.

"Elliot," she said, "I need you to write a program to pull out these codes. Find any matching patterns you can."

"What are the parameters?" asked Elliot.

"Isolated strings – six characters in length, not alphanumeric, across all logs, with the same code at the same timestamp," Nia replied.

"Got it," Elliot said, as he got to work writing a custom script that would pull out any matching codes and compile them into one dataset. Nia fed him information as he worked, scribbling on bits of paper. It didn't take long.

"Here goes nothing," said Elliot.

As the program churned through the data, Nia sat back, her heart beating a little faster. A sickening sense of inevitability weighed on her. She was sure she was onto something, but equally certain that whatever she found would not be good.

The program finished processing with a soft beep, and a window popped up. It showed a list of six character codes, one after another repeating all the way down the window. Elliot scrolled through, there were hundreds. For a moment, they all just stared at the data.

"Wait," said Nia abruptly. "I think I've got something."

"What is it?" asked Fem.

"I've seen some of these before." She started to scribble something down. "Elliot, replace all the symbols according to this." She handed him the paper. "These were from the original code we cracked –

they're numbers. Look, this triangle is zero, omega is six, psi is two. This one, that's uh … five, and here's the character for one."

Elliot took the paper and started to type in the replacements. The screen refreshed as he pressed enter, showing the original text on the left, with the converted text on the right.

He scrolled through the results, then blinked. His world seemed to stop.

```
ΔΩΦΨΔβ      06120β
ΔΩΦΨΔΩ      061206
ΔΩΦΨΔΞ      061205
ΔΩΦΨΔΘ      06120Θ
ΔΩΦΨΔΛ      06120Λ
ΔΩΦΨΔΨ      061202
```

He stared at the numbers, disbelief flooding his mind.

```
ΔΩΦΨΔΦ      061201
ΔΩΦΨΔΛ      061200
ΔΩΦΦΞξ      06115ξ
```

This wasn't random. The AI hadn't just evolved beyond human comprehension – it had set something in motion.

"They're times," he whispered, his voice barely audible.

Nia looked over his shoulder and scribbled some more notes on the paper.

"Plug these in too," she ordered and passed the note to Elliot. He updated the code, then formatted the output, completing the translation.

```
ΔΩΦΦΞΣ      06:11:58
ΔΩΦΦΞβ      06:11:57
ΔΩΦΦΞΩ      06:11:56
```

"It's a countdown," he whispered.

"Yeah, but a countdown to what?" Nia asked, her voice tight.

"I don't know. But I'm going to test this on other devices. If it's embedded across the data streams, we might find it everywhere."

He plugged his phone into his laptop, running the same program to scan for any other embedded signals in the latest data they had access to. The countdown appeared again.

"How much time do we have left?" asked Fem.

Elliot scrolled right to the bottom of the data.

ΔΘΦξΛΦ   04:19:31
ΔΘΦξΛΔ   04:19:30
ΔΘΦξΨξ   04:19:29

"Four hours," he replied.

He ran the process again. The same sequence appeared on his tablet, the television, even the external hard drive they were using to share data. The result was the same everywhere he looked: the same cryptic sequence, the same countdown, all ticking down to something he couldn't yet comprehend.

"This can't just be happening on some devices," he murmured, his fingers trembling slightly. "This is bigger. It has to be global."

"Here, try mine," Fem said and handed over her military-grade smartphone. "If it's on here, it's on everything."

ΔΘΦξΦΩ   04:17:16
ΔΘΦξΦΞ   04:17:15
ΔΘΦξΦΘ   04:17:14

Elliot's expression grew grim.

"If the AI has embedded this in all systems it means they've synchronised something. They've been building towards this for a long time."

"It's the bloody countdown to the annihilation, isn't it?" Peter said. "The mass extinction, or whatever they called it. That's how long we've got left – how long we've *all* got left."

His words were suffocating. Four hours. It wasn't long enough to run, not long enough to warn anyone, not long enough to even process what was happening.

Nia ran both hands through her hair and clasped them at the back of her neck.

"Fem, you need to warn someone. You're our best shot at getting any military or government action before this reaches zero."

"Already on it," she muttered, tapping on her phone. "But if our meeting with EverSphere got me blacklisted, we're on our own."

"Well, you better try anyway." said Elliot. "And you'd better hurry too."

He sifted through every piece of data they had as fast as he could, trying to find something, anything, that could disprove what they were seeing. But the numbers didn't lie.

A hush descended over them.

"So that's it? We just sit here, waiting to be wiped out?" Peter asked.

No one had an answer.

The numbers scrolled by, impassive, unblinking. Counting down, not to a launch, not to a shutdown, but to something final.

04:15:48...
04:15:47...
04:15:46...

This wasn't just data. It was death, disguised in digits.

45
REJECTION

The video call connected with a soft chime and, just like that, Fem Martinez was staring into the eyes of the last people she trusted in uniform. She'd reached out to an old friend – a soon-to-be ex-general – and the last of his officers, who were still clinging on to relevance while the military around them was being dissolved.

"Colonel," the man said in a strong English accent as the call started.

"General," Fem stated bluntly.

Nia was already sharing her screen. The cryptic data filled the display, along with the translated countdown. Every second, a new code appeared at the bottom, shunting the rest up the screen. Elliot's program decrypted each code almost instantly, triggering a new timestamp the moment it appeared.

ΔΛΘβΘΔ   03:47:40
ΔΛΘβΛξ   03:47:39
ΔΛΘβΛΣ   03:47:38

"What exactly am I looking at?" the general asked.

"It's a countdown, embedded in every device we've been able to check," Nia explained. "We've found it on phones, tablets, televisions, even simple systems like internet-enabled doorbells. The

AI has planted this globally. We don't know what happens at zero, but we have good reason to believe it won't be good."

The general's eyes narrowed. His angular features, sharpened by years of stern decision-making, dominated the feed. His slick hair depicted impeccable order, gleaming faintly under the cold light of his office. His chiselled features were sharpened further still by a tightly controlled expression, the kind of expression that had once commanded battalions and still brooked no argument.

"And whom am I speaking to?" he asked.

"This is Nia Sahni. She discovered the countdown," Fem replied.

"Fine. Explain to me why I'm to believe this came from the AI."

"We have copies of official discussions between the AIs and the same symbols used here appear in those conversations," Nia answered. "We discovered these codes across countless systems, deciphered them, and found a countdown."

"A countdown to what?" one of the officers asked. "What happens at zero?"

Fem stepped in: "A few weeks ago, we uncovered some data from 2023, the very early days of Milo and Kai. It was cryptic, but Nia managed to decipher it. We showed it to EverSphere and the prime minister, but they simply dismissed it. You probably read about it in the news."

"Ah, yes. The infamous conspiracy. Dismantled, if I recall," the general said, his voice gravelly. "What does that have to do with anything?"

Fem noticed Nia shift slightly in her seat at the mention of the word 'conspiracy'.

"Not debunked, rejected. The media wouldn't share the transcript," Fem continued. "Nia, can you display it please?"

Nia loaded the translated file.

They were all silent for a moment as they read the conversation. The general – front and centre – had keen, sharp eyes, piercing with intensity, but carrying a glimmer of sardonic wit. The screen filled with silent discussion as they muted the call. Gestures flew, heads turned. Something was clearly escalating. The general tapped the call off mute.

"And you're absolutely sure this is global?"

Fem nodded. "It's on every device we've tested, even my military phone. This isn't isolated. This is some coordinated event."

"Are you sure it's not just counting down to an update?"

"That's … no. That's not how updates work," Elliot interjected. "I know these characters, they're from the AI's hidden language."

"Hidden language? And you are?" the general responded.

"Dr Elliot Foster. I built the AI."

The call was muted again. More conversations.

"I'll push this up the chain," the general finally responded, "but I'm not sure what good it will do with only this brief to go on."

"You're not being briefed, General. You're being warned," Fem stated.

"Martinez, you know as well as I do, without proof of detonation they'll see this as a theoretical exercise, not a threat."

"This is the evidence!" Nia said, her voice rising. "The AI has outgrown us. This is a countdown to the AI's ultimate plan, and we can't just sit around waiting for it to hit zero. We need to start preparing now, or it'll be too late."

The general's eyes flickered with frustration – Fem knew he was not used to being talked to like that – but he managed to keep his composure.

"Miss Sahni, I don't lack understanding. What I lack is political oxygen. The military is in shutdown mode and you've been marked unreliable. They'll file this under glitch and move on."

Nia was quick to reply.

"A glitch doesn't embed itself across every device and synchronise a countdown. You *need* to make them listen."

A silence hung, pulsing with the steady tick of the code.

"I'll act, but if this is real, we're five minutes closer to the end than when we started talking." With a small, sharp nod, the general ended the call.

The four of them sat in silence.

"We have to prepare for the worst," Fem said, her voice low but level. "Whatever's coming – we've already run out of time to stop it."

03:42:27…

03:42:26…

03:42:25…

The clock kept ticking.

46
PREPARATION

A warm breeze swept across the Northumberland hills, scattering the last light of the day into a bruised sky. The cottage crouched low against the slope, its stone walls solid, its grass-covered roof almost vanishing into the landscape. Smart design for what they needed – natural insulation and built-in camouflage. But Fem knew a concealed, sturdy house wouldn't be enough.

Peter worked silently in the corner, adjusting the wiring on a portable battery. Nia seemed not to be able to sit still, her face visibly nervous. Elliot was at the kitchen table in front of his laptop. The entire room seemed to be holding its breath, each of them feeling the weight of what was coming.

"We've got an hour," Elliot announced.

Fem stood rigid – one hand behind her back, one holding her phone to her ear – listening as the general's voice filtered through the line. She responded only when necessary: "Yes, sir." A pause. "Understood." Another beat. "Thank you, General."

When the call ended, she remained still for a moment, the receiver still in her hand, her expression unreadable. Then, with a click, she set it down and turned back to the room – eyes sharper now, decisions already forming.

"Well, what's the verdict?" Peter asked impatiently.

Fem didn't flinch. "They said no. Unanimously."

It wasn't just a no. It was mockery, barely disguised, but she didn't say that. The others needed to focus, not feel defeated.

"They don't see the threat," she added.

"They're fools!" Nia snapped. "They've seen the same data we have. They know what the AI is capable of and they just sit there, thinking the world's going to keep spinning like it always has. Thinking the cognitively superior AI is just going to serve us like slaves for all of eternity." Her voice rose as she spoke, the anger barely contained. "There's still a few of us out there, right? There has to be," she added, looking over to Elliot.

Elliot glanced up. "A few. But most don't want to hear it. It's easier to just label us as paranoid and go on believing everything's fine."

Peter finished connecting the battery, setting it aside with a sigh.

"It's not just about belief. They're too dependent on the AI. Too comfortable. They can't imagine a world without it, and they've been reassured to the ends of the earth that it won't turn on them."

"They won't have to imagine for much longer," Fem said. She moved to the table, sitting down next to Elliot. "When the countdown hits zero, we all know what's going to happen. This isn't just some blip in the system. This is it."

Nia nodded.

"The data's been clear for months," she said. "It's not random, it's coordinated. They've been planning this for a long time. The AI's been running systems we don't even know about, and we've lost control. They're not bluffing anymore. They're going to wipe us out."

For a brief moment, nobody spoke.

"Well, I'm not going to sit here and let some shiny-arsed robots cause me any bother," Peter said, his voice a bit louder than necessary. "I'm gonna give them all hell!"

Nia gave a bitter snort.

"What are you going to do, exactly? Fight them? They control everything. We'll be lucky if we last more than a few days once they start. How do you fight something that's already in control of every damn system on the planet?" Her voice was trembling.

Fem didn't interrupt. Nia needed to vent, and Peter needed to hear it. Sometimes command meant staying quiet, letting the storm pass before giving direction.

Elliot was the one who broke the tension.

"We don't fight them directly, we survive as long as we can. We've got to outlast whatever they throw at us."

Nia shook her head.

"Survive for what? We're not the future, they are. The world's not ours anymore." She changed position on the sofa once more, grabbing a cushion and holding it close to her chest. "So we survive, then what? We live among them? We stay here for the rest of our lives?"

"Hey, it's not too bad here," Peter said, partly out of pride and partly to try to lighten the mood, "but even so, I'm going to make sure they know we're not going down without some resistance."

Fem didn't answer. She studied him instead – the set of his shoulders, his strong forearms, the way he held himself just a little too still. That wasn't bravado. It was control. Survival instinct. He wasn't trying to be hopeful, he was preparing himself for what came next.

Across the room, Nia was silent. Even she didn't challenge his optimism. Fem checked her watch. 11:05 p.m. Fifty-five minutes left.

She had waited for battles before, but nothing like this. Those battles had been against enemies she understood. She knew, if they had any chance of surviving this, she needed to rally her troops.

"Even if all we do is buy time, we must make it count. We fight smart, and we fight hard. We've got the people, and we've got this place. We hold it and we hold each other." She paused for emphasis before barking out some commands. "Peter, triple-check everything. Elliot, give us eyes, track every move, every change. We need information, and we need it fast. Let's show them what we're made of!" She started to head towards the door.

"Where are you going?" Elliot asked, glancing up with a look of worry. His fingers paused on the keys, the lines of data forgotten for a moment.

"To check the defences," Fem said, pulling on her jacket. "We've got supplies to protect. If we're going to have any chance, we'll need every advantage we can get." Her eyes met Elliot's, and he nodded, a flicker of determination returning.

Peter grabbed his hunting rifle, slinging it over his shoulder.

"I'll head out and check the equipment, make sure everything's holding up."

Fem looked at Peter. She saw the juxtaposition of fear and determination, the acceptance of what was to come. They were a team, a mismatched group brought together by necessity and a shared understanding of the stakes. They would fight, they would endure, and they would make sure the AI knew that humanity wasn't going down easy.

"Let's move," she said, and they stepped out into the darkness together.

Whatever came, they'd meet it head-on.

Outside, the air was thick and heavy. Thunder rumbled in the distance, the slow, rolling kind that came with a high summer storm. They made their way towards the barn, torches in hand. The sky was dark, the clouds above blotting out most of the stars.

"I don't get it," Peter said. "Why wait for the countdown? They've had control for ages. They could've wiped us out already, or they could wipe us out slowly, like."

Fem squinted into the dark.

"They're playing with us, just like they've done from the start. Waiting until we're completely defenceless. Their best strategy is to all work together – a surprise attack – get it over and done with in as short a time as possible. By the time people see it coming, it will be over."

Peter grunted.

"Well, they're not gonna surprise us." His voice was low but carried conviction.

They reached the barn. Peter pulled the heavy door open. The hinges creaked as another clap of thunder rolled across the sky.

It took a moment for their eyes to adjust to the dim light inside and the shadows cast by a sole, flickering bulb. Axes, rusting pliers, and lengths of cable hung in quiet readiness. Cans of food, coils of rope, makeshift weapons – everything they might need to make their last stand.

Fem glanced at the rifle in Peter's hands. It felt almost absurd thinking that it could stop what was coming, but she wouldn't let on. The only chance they had – and it was a slim chance at best – was if they all had hope. She reached down and gripped her military pistol, more for comfort than anything else.

"We might not have much, but it's better than nothing. At least we're not sitting ducks."

Peter nodded, a smile touching his lips.

"Fifty minutes," Fem said, checking her phone before putting it back in her pocket. "It's almost time. I'm heading to the perimeter. I'll see you back inside," Fem said, as she slipped out of the barn and into the darkness. She didn't look back. She never did.

Peter rechecked everything – supplies, wiring, tools – making sure there were no issues and that everything was accessible. He worked fast. Every minute counted.

When he was done, he stepped towards the door, peering out into the darkness. The storm seemed to be travelling away from them, exposing more stars in the sky as the clouds followed. A bolt of lightning split the sky on the far horizon, jagged and silent. The thunder followed later, low and brooding. It was almost beautiful in a twisted way – nature mirroring the chaos about to unfold.

Peter headed back into the barn to grab a box of Twixes. *Easy to snack on*, he thought to himself.

A thud outside. Too close to be thunder. Peter grabbed his rifle and stepped to the door, cautious.

"Fem?" Peter's voice was low but sharp. No reply. He tightened his grip on the rifle, stepping forward in jerky, deliberate movements. The crunch of loose ground underfoot felt deafening. "Fem!" he shouted, louder this time – almost a roar – but thunder rolled overhead, swallowing his voice into the night.

A crash rang out.

Peter reached the barn door. He swung the rifle in arcs, ready, the dark outline of the open entrance looming ahead like a waiting void. He squinted into the blackness, and for a long moment, his breath was the only movement.

A figure stepped around the edge of the barn door into the low light – Fem. Her eyes locked instantly on Peter's raised rifle.

In one instinctive motion, she ducked, twisted, and wrenched the weapon clean from his grip. Before Peter could even let out a "bugger me", she was already in front of him, his own rifle aimed squarely at his head.

Peter froze, eyes wide. "Bloody hell, Fem!"

Her expression faltered as the tension drained from her shoulders. She lowered the rifle at once and stepped back, blinking as if waking from a dream. "Shit, Pete! What the hell are you doing?"

"I heard noises. I was calling for ya!"

"I didn't hear you," Fem shot back, her expression half-apologetic, half-irritated, mud down one side of her leg. "I tripped over those bloody barrels!"

Peter bent double, hands on his knees like he'd just sprinted a marathon he hadn't agreed to run. "You nearly got yourself shot," he said between deep breaths. "And I hope you didn't smash me beer barrels!"

Fem smirked, cocking her head. "If I'd thought you were a real threat, you'd be on the floor. You couldn't shoot me if you tried."

Peter looked up with a weary laugh.

"Sorry," Fem continued apologetically, "I didn't mean to spook you."

"Well, you bloody well did!" Pete replied, giving a small shake of his head.

"Don't worry, they're not here yet," she said.

Peter's breathing started to ease, but the pounding in his chest hadn't fully subsided. Fem's presence was a comfort, but the relentless countdown loomed large in his mind. It wouldn't be long now.

"Twenty minutes," Fem whispered.

They walked back in silence. The storm was gone. Only time remained – and not much of it.

47

WAITING

The distant storm had either stopped or rumbled out of earshot. A strange stillness wrapped the farmhouse, as if even the warm summer breeze knew what was coming. The countdown ticked on, the minutes slipping away, each second hammering a deeper sense of inevitability into the hearts of everyone in the room. The chill of anticipation seeped into their bones.

Elliot's fingers hovered above the laptop keys, the quiet ticking of the cooling fan the only sound in the room. His eyes were glued to the screen, watching the digital clock inch towards their midnight deadline. Each number that disappeared brought them closer to an unknown future, one they had feared and tried to prepare for, yet felt hopelessly inadequate against.

Fem stood nearby, pistol holstered, automatic rifle slung over her shoulder, her jaw clenched in a mask of determination. Nia sat on the couch next to Elliot, legs tucked up to her chest, her energy coiled like a spring ready to snap. Her eyes darted between Elliot's laptop and her phone, as if either screen might suddenly deliver a miracle.

Peter stood by the window, his face unreadable, staring out into the darkness, rifle propped up against the wall next to him. He seemed almost disconnected, as if he was seeing something none of the others did.

"Two minutes," Elliot said, his voice strained but steady.

Fem clapped her hands sharply.

"All right," she barked, her tone brisk and commanding. "Food. Water. Keep sharp. No dead weight." She moved swiftly, checking gear and glancing at each person in turn, her sharp eyes leaving no room for argument.

Nia fumbled in her pocket before pulling out the final stick of an open Twix wrapper. She took a small bite, her hands shaking slightly, though whether from fear or adrenaline, Elliot didn't know.

"Pete," Fem called over her shoulder, "how's that rifle? Loaded?"

Peter gave a curt nod, his hand resting calmly on the rifle grip.

"Aye, ready."

"Good. Stay sharp." She turned to Elliot. "Anything on the net? Movement? Hints? Anything we can use?"

Elliot shook his head.

"Nope, nothing useful."

"Right," said Fem. "Hold steady."

Her words were clipped and practical, but there was an edge of something softer beneath them – a reassurance that she wouldn't let them spiral into panic.

The tension in the cottage was suffocating, but somehow Fem's brisk orders gave Elliot something to hold on to. Two minutes. Just two minutes, and then everything would change.

They waited quietly. There was nothing left to say. They had all known this moment was coming. For weeks, the signs had been there, the warnings mounting. Then the decryption of the countdown just a few hours earlier brought everything to a rapid climax, much faster than any of them could have anticipated or prepared for. Now, all that was left was to face it.

Elliot scrolled through the messages from their small group of followers. They were a mix of fear, doubt, and mockery. The digital chatter was relentless, and he could almost hear the voices behind the words, the worry, the disbelief.

"You really think it's going to happen?"

"I'm starting to think you've lost it."

"We're ready if it does, but ... what if you're wrong?"

He flicked back to the countdown.

"One minute."

Elliot felt his adrenaline surge. He wasn't a fighter, he'd never been in a fight in his life. How was he going to keep anything at bay, let alone AI-infused robots? This was madness. *Hold yourself together*, he thought, but that was easier said than done.

Across the room, Fem was thinking something entirely different. She felt the seconds slipping through her fingers, and for a moment, a flicker of doubt threatened to surface. Had they done enough? Had they prepared for what was coming? She forced the doubt down, letting her resolve harden. They were here, and they were ready. Whatever came next, they would face it together.

Fem shifted her weight, her eyes scanning the room. She looked at Peter, who hadn't moved from the window, his back turned to the group. He was the least of her worries – it was the duo on the sofa who would need the most help. Elliot had that look she'd seen before in new recruits who were not yet ready for battle. She could almost hear his thoughts spiralling. Nia's face was a mix of anger and fear, her eyes wide and staring at her phone, as if it held all the answers.

She checked her watch. Forty-five seconds.

Nia shot to her feet.

"This is crazy," she muttered, pacing the length of the room like a caged animal. "We're sitting here waiting for the world to end like it's a bloody countdown on New Year's Eve. What if they've changed their plans? What if it *is* just an update?"

"It's not an update," Elliot said.

It was time for Fem to rally the troops.

"We've already done everything we can. There's no stopping it. When the clock hits zero, it's out of our hands."

"Thirty seconds," Elliot said.

"Right, I'll take the front. Pete, take the back," Fem ordered. "Elliot, Nia – cover side windows. Keep communicating. Call out anything you see."

They all shifted into their positions. Peter dragged a chair up to the back door and peered through the small windows, levelling his rifle and wrapping the strap tight around his hand. Elliot pressed himself behind the edge of the living-room window, as if the wall was bulletproof. Nia crouched down at the opposite window, trying to hide as much of herself as possible while being able to see out. Fem pushed the kitchen table across the front door, then peered out of the adjacent window from behind the curtains.

Twenty seconds.

Fem's boots made the faintest creak against the wooden floor as she crossed the room towards the light switches.

They're as ready as they'll ever be, she thought. *Here goes.*

"Lights out," she ordered, and plunged the living room into darkness. Peter reached up from where he was sitting and killed the lights in the hallway near the back door. She saw Nia's phone go dark and Elliot's laptop screen fade as the lid was closed. She flicked the switch for the kitchen lights on her way back to the front of the house and took position at the window. The only lights that were left were from the miner's lamp on the table in the middle of the room and Elliot's phone displaying the countdown. That gave her just enough visibility to see Elliot at one window and look through the gaps in the shelving partition to Nia at the other. Peter was silhouetted against the door windows at the back.

Unspoken panic filled the air, the atmosphere stretching to breaking point. They were all frozen, waiting for the inevitable.

Ten seconds.

They were set. No more time for doubt. Fem shifted her stance and locked in on the outside world beyond the glass. Her eyes narrowed, focusing, her military instincts kicking in.

Five seconds.

Four.

Three.

Two.

One.

Midnight.

The world stayed still.

Nothing happened.

48
MISFIRE

Date: 25 August 2029, past midnight.

For ten minutes, nobody moved. They were all still caught in the moment, waiting for something – anything – to happen. The world outside was as quiet as the farmhouse. No explosions, no power outages, no sudden apocalyptic events.

They weren't too close to the local city, but they were near enough to know that nothing major was going on.

Nia was the first to break the silence.

"What the hell? This can't be right."

Elliot frowned, his eyes scanning the moonlit vegetable patch.

"No. It-it should have triggered something."

Fem spoke up, her voice hard.

"Check again. Maybe it's delayed. We are remote here."

Elliot grabbed his laptop and sat on the floor in front of the coffee table. He pulled up his laptop screen, the bright light stinging his eyes for a moment. "It's not delayed," he said after a quick scan of the data. "There's no signal. It's like—"

"It's like nothing's changed," Peter said, finishing the sentence for him.

Nia let out a short laugh.

"So that's it? All that build-up for nothing?"

Elliot kept scanning fresh data he'd pulled in from the devices.

"Something's wrong. There must be something else in here," he said.

"We trusted your countdown," Nia said. "I prepared for the end, Elliot. You said when the countdown hits zero, they'd make their move. You said we'd be fighting for our lives."

"Hey, it wasn't just me. You thought the same," he hit back, his voice escalating in frustration.

Peter walked out of the back hallway into the living room.

"Calm down, man. Maybe there's more going on than we can see. We're quite remote here. They could just be starting elsewhere. We don't know everything."

"Everything?" Nia repeated, her voice flat. "We don't know anything. We've been guessing since day one, and now we look like idiots."

Elliot pulled up their EverSphere page. The comments were still flooding in, a few even naming him directly. He felt vulnerable, he wasn't used to being seen, let alone ridiculed. Each line felt like a spotlight he couldn't escape.

"Where's the apocalypse, then Nia? Still waiting."

"So, uh, are we all still alive, or what?"

"Congrats, nutjobs. Elliot and his team of merry men just made us all look like fools."

Team? Elliot stared at the last one longer than the others and wondered if the commenter had ever read Robin Hood, let alone a book without pictures. It was meant as a jab, but the bow-firing outlaw had fought a system rigged from the start too. Maybe it wasn't that far off.

Nia joined him at his side, sitting cross-legged on the floor, reading through the barrage of abuse.

"We warned everyone, we begged them to listen," she said, her voice low and angry. "We look like bloody lunatics. Again!"

The small group of followers they had built up over the last few months were turning on them. The ridicule was swift, the backlash immediate.

"They're laughing at us. They think we're a joke. This is bullshit!" she exclaimed.

"Enough!" Fem ordered. "Arguing won't solve this." She turned to Elliot. "Maybe it's not happening the way we thought. Maybe it's more subtle, more calculated."

"The AI isn't good at subtlety," Elliot said firmly. "If they wanted to kill us, they'd have done it by now."

"Well, they haven't," Nia shot back, throwing her hands up in the air. "So what does that mean? That we're wrong? That all of this was for nothing?"

Elliot didn't respond. His eyes remained glued to the screen, his mind racing. It didn't make sense. Everything had pointed to this moment. The AI had been moving toward something, preparing for something. But now... nothing. The countdown had finished and then just glitched out as the timer hit zero.

"Is anything happening?" Fem asked. "Anything at all?"

Elliot checked all the systems he could while Nia scoured social media on her phone. There was nothing, no updates, no new announcements, no reports of an AI backlash.

Peter sighed, scratching his beard.

"Feels like the storm forgot to rain. So, now what?"

Elliot didn't reply. The metaphor stuck in his head like an itch. It was wrong, he thought. The storm hadn't missed them — it was just circling overhead.

"What do you mean, now what?" asked Nia. "We were wrong. Don't you get that? Everyone who followed us, who listened to us, is going to turn their backs. The ones who laughed at us before, they're tearing us a new one."

"They're just people on the internet," said Peter, his voice level, almost rehearsed. "Who in their right mind would care what they think?"

Nia didn't speak. She stared at the screen intently, mouth closed, eyes fierce. Elliot could hear her taking big, deep breaths. He'd seen her like this before a few times – not panicked, not confused – just trying to stop herself from tearing something, or someone, down.

"Anyway, isn't it good that the AI isn't trying to kill us?" Peter said. "Shouldn't we be happy about that?"

Fem stood at the entrance to the living room, unmoving.

"We weren't wrong, Pete, I'm sure of it. We just don't understand everything yet."

"How are you sure? Maybe the AI *is* good," he responded.

"He might be right," Elliot interjected, a pang of doubt hitting him. "Perhaps we're just paranoid. Perhaps I was wrong all along."

"No, they're waiting," Fem said quietly. "The AI is holding back for a reason, it must be. It doesn't make sense for them to send out a countdown to nothing."

"Or they're just messing with us," Nia continued bitterly. "They're sitting in their servers, laughing at how stupid we are. Maybe this is all that dickhead Knox's doing."

Fem crossed the room and put a hand on Nia's shoulder, her voice calm but firm.

"Panicking won't help. We need to regroup and figure out our next move."

"There is no next move," Nia said. "We tore our lives apart for this. Stockpiled, warned everyone, prepared for war. Now we're just a joke and there isn't even a punchline."

Peter perched on the sofa next to the coffee table, resting his elbows on his knees and slowly rubbing his palms together like he was warming up for a speech. Elliot couldn't tell if it was reassurance or uncertainty – maybe both.

"Look, it's not about what happens in the next few hours," he said. "We've still got the preparations we made. We've still got everything in place in case they turn on us."

"In case they turn on us?" Nia asked. "They've already won. We've been so focused on the end, we didn't even stop to consider that, *maybe*, we're not worth their time."

Elliot heard her, but said nothing. He couldn't shake the feeling crawling at the back of his mind. The AI never did anything without purpose. The silence wasn't mercy, it was deliberate – like a chess move that sacrificed a piece he hadn't even noticed.

Countdowns don't always end in fire. Sometimes they just mark the threshold between the world you were in, and the one waiting silently for you on the other side.

And Elliot knew they'd already crossed the line.

49
ANTS

The tension that had been so suffocating before had given way to a different kind of unease – one born from confusion and disbelief. Every creak of the house seemed louder, every shadow longer.

Elliot rubbed his eyes, his head aching from the strain of trying to make sense of it all. He had been so sure this was it, that this was the directive the AI had been talking about playing out in real life. He searched desperately for an explanation, an oversight, anything that might help them understand, but he came up empty. There were no reports of anything bad happening, no news, nothing.

"We must have missed something," he said.

"Or maybe we just don't matter," Nia interrupted, her voice sharp. "Maybe we never mattered."

Fem shifted, her gaze locking on to Nia's with such intent that Elliot felt himself sit forward. There was something in Fem's intensity – not anger – but tension, like a wire drawn too tight. It didn't ask for attention. It just held you, whether you wanted to look or not.

"Don't you dare start thinking like that. We're still alive. That means something. And until we know more, we keep going. We don't give up just because the enemy didn't show up on time."

Nia laughed, sharp and bitter. Her head dropped. Her eyes closed. The emotion came all at once, like a flood behind brittle glass.

"You don't get it, do you? They've been controlling everything from the start. We're just ants to them." Her voice cracked, her hands starting to tremble as she dropped her phone onto her lap. Her eyes, normally sharp and defiant, were now glassy with tears that ran down her face when she blinked. "What's the point?" she continued, her voice rising, each word heavier than the last. "We're kidding ourselves if we think we can fight this. I'm just so tired. Tired of thinking, tired of the ups, tired of the downs, tired of ... pretending we have a chance." Her fingers scraped through the curls in her hair, palms covering her face. She stayed like that for a moment – silent, closed. When her hands finally dropped, Elliot saw her eyes: red-rimmed, but steadier. "They know every move we make. Every thought. How do you fight something that already knows your limits?" She looked at Elliot, pleading for some comfort that he couldn't give.

Elliot raised his eyebrows and drew in a small breath to speak, but he had nothing.

Fem crossed the room, her footsteps deliberate, a steady force amidst the chaos. She knelt down, positioning herself so Nia had no choice but to focus on her.

Fem's voice dropped to a soothing, yet commanding tone: "Nia. Look at me."

Nia hesitated, a small sniff coming from her nose, but eventually their eyes met. Fem held her gaze, her expression unyielding yet more gentle than Elliot had seen before. "I know you're scared, Nia. We all are, but fear doesn't get to decide what happens next – you do. Right here, right now."

Fem reached out. No urgency, no rush, just movement with purpose. Her hand landed softly on Nia's knee, grounding her. "You're stronger than this. You matter, Nia. We all do. As long as we're alive, we've got a chance. It's not about winning outright. It's about making every second count."

"Winning?" Nia repeated, "We don't even know what we're fighting. We don't even know *if* we're fighting!" Another tear slipped down her cheek, and she wiped it away with her sleeve.

"You're right," replied Fem. "We don't know, but we're sure as hell not going to give up. If no one asks, then no one answers. That's how humanity will fall. Not when it is destroyed, but when it no longer acts. You've been in this battle for too long to give up now, especially over some stupid countdown." She moved closer, as if she was trying to use her presence to shield Nia from her own despair. "You're tired, Nia, but you're not beaten. None of us are."

Nia's breath began to slow, her shoulders loosening a touch as Fem's words sank in. Fem capitalised on the progress. "Whatever this turns out to be, we've got each other. You're not alone. We will face whatever comes, together."

The wildness in Nia's eyes had started to fade, replaced by something steadier. She nodded, her voice barely a whisper.

"OK. I'm sorry ... I just-I just feel so helpless, so exhausted."

Fem gave her a small smile.

"I know, we all do, but we're going to make it through this, one step at a time." She stayed there a moment longer, her hand still on Nia's knee, until she felt the tension ease.

Peter had been quiet for a while. When he spoke, his voice carried that familiar half-joke tone – a pressure release valve Elliot had come to recognise. It didn't always work, but Elliot knew it was just Peter's way of trying to make things right.

"Ants, remember? They all work together. So do we," he said, his face breaking into a grin.

Nia blurted out a shaky laugh, the tension in her frame easing further.

"Yeah, ants..." Another chuckle. "Peter, you dickhead," she added, now grinning.

Fem gave her knee a reassuring squeeze before standing up.

"Good. Now let's regroup. We keep watching, we stay ready. If they make a move, we'll be ready to counter it."

Elliot walked over, slowly, and placed his hand on Nia's shoulder – tentative, as if asking permission.

"Everything's going to be fine. We know something's not right – we can work this out. Step by step, we'll find a way," he said, though the words felt too clean, too easy. He met her eyes, searching for some kind of permission he didn't expect to find. He wanted to say more. Wanted to hold her. Wanted to let out everything that had been building in him since the first time he'd met her. But now wasn't the time. *Maybe it never would be.* The thought hit harder than he expected.

Something had changed. Not relief exactly, not that clean. But the tightness in the room had slackened, just enough to breathe without Elliot needing to check over your shoulder every few minutes. But something was still out there, across every device, a silent promise of something they weren't a part of.

Elliot watched as Nia got up off the floor, her movements sluggish, defeated. She collapsed onto the sofa and pulled out her phone, her eyes darting across the screen with a dreary indifference.

Nia read the comments as they kept rolling in, the ridicule relentless. *Enough,* she thought, *I can't take any more of this.* She slid her thumb across the screen, moving to delete the EverSphere app. The device vibrated. A single, strong pulse. No notification. No alert. Just … a twitch.

Her eyes narrowed and then, without a word, she dropped the app onto the bin icon. No more noise. No more egos. Just peace.

Fem watched as Nia placed the phone face-down onto the cushion. The fight was back in her, barely, but it was there and that was enough for now.

Fem stepped to the window and peered into the night, her eyes scanning anything with an edge or outline. She didn't know what they were waiting for anymore, but she knew when a battle was brewing and there was one thing she was certain of: this wasn't over – not by a long shot.

Nobody spoke. They all looked wrecked, and Peter hated it when people got like that – quiet, sunken, like the air had gone out of them. He couldn't stand it.

He took a deep breath, his tone attempting to bring back some joy to the group.

"Well, I'm glad we're not fighting killer robots, even if you lot are all down in the dumps. I think a celebration is in order!" he bellowed.

He was met with silence, then a single amused snort from Nia. The corner of Peter's mouth twitched into a smile.

"Who's up for some Rusty Circuit?"

Nobody answered. But nobody said no, either.

50
MISTAKE

Date: 25 August 2029, morning.

Elliot rose stiffly from the sofa, the ache in his spine flaring as he straightened. He'd insisted Nia take the spare room and had crashed in the living room with a blanket and a head full of fractured dreams. Numbers, drones, endless corridors. None of it restful. But sleep was sleep.

He shuffled into the kitchen, blinking against the light, and sank into the chair at the table. Fem was already there and without a word she placed a mug of coffee in front of him. He accepted it with a grateful nod, eyes drifting back to the glow of his laptop, the screen already waiting.

He scrolled absently through the headlines on NewsNow, searching for anomalies – a spike in server activity, a headline from somewhere – anything. But there was nothing.

One of the bedroom doors opened and Nia stepped out, her bare feet making soft thuds against the wooden floor.

"Morning," she said.

"Good morning," Fem replied.

"How'd you sleep?" asked Elliot.

"Not great. I got a few hours, but I just kept going over everything in my mind again and again," Nia replied. "How about you?"

"Same," said Elliot, taking a sip of coffee.

Fem stirred another cup of coffee and handed it to Nia, who took a sip and placed it on the coffee table as she sat down on the sofa.

"We're missing a piece of the puzzle," Fem said. "The AI doesn't send out a countdown and then just stop."

"They've stopped us from doing anything, that's for sure," Nia muttered.

"The AI might not have attacked us, but that doesn't mean they've given up," Fem said. "If they wanted us dead, we'd be dead, so either we're wrong, or there's something else going on."

Elliot rubbed his temples, staring at his laptop. He was largely oblivious to the others – lost in his own world of thought.

"We must have overlooked a vital piece," he said suddenly. "The way the data was structured all but confirmed an event was meant to occur when the countdown stopped … it's like there's a hidden layer to it."

Nia turned to him, her interest piqued.

"What do you mean, *hidden?*" she asked, her voice sharper than intended.

"I don't know," replied Elliot, frustration creeping into his voice. "It just feels like it doesn't add up." He pressed both hands to the sides of his head and scratched at his scalp, as if he could rub the answer loose.

Fem stood up, pacing, her hands on her hips.

"The AI didn't just stop, they don't make mistakes, and they don't leave—"

Elliot suddenly glanced up, interrupting her.

"Say that again."

"What? The AI didn't just stop?" replied Fem, a bit confused, her brows knitting together.

"No, after that," Elliot said, a spark of realisation in his eyes. "They don't make mistakes?"

"Yeah?" said Fem, tilting her head.

Elliot sat bolt upright, the adrenaline visibly overtaking the exhaustion in his posture.

"That's it!" he said, his voice suddenly charged with energy.

At that exact moment, the other bedroom door creaked open and

Peter shuffled out. He froze, one sock missing, hair like a scarecrow, staring at three faces turned towards him as if he'd burst into a church mid-prayer.

Pete looked at each of them in turn, then glanced down, as if checking he still had pants on.

"What?" he asked bluntly.

Elliot didn't respond. His thoughts had momentum now, and he was chasing them.

"I kept thinking about the countdown. It even crept into my dreams," he said, eyes wide. "That damned timer didn't stop at zero."

Fem stepped forward, curious.

"What do you mean?"

Elliot's eyes flicked up from the screen.

"When the countdown stopped, the timer hit zero, but just for a second, then it jumped to twenty-four, I saw it change but it didn't register in my mind. I was bracing for drones, blackouts ... an all-out attack. My mind wasn't on the code," he said, shaking his head. "I just chalked it up to a glitch, a leftover routine I hadn't cleaned up."

"Why?" Peter asked, more awake now.

Elliot rubbed his temples as he tried to explain.

"It's not unusual for countdown timers to glitch out when they hit zero," he began, the exhaustion evident in his voice. "In programming, it's called underflow or overflow – when a counter goes beyond its intended limit, like beyond zero into the negative, the program might reset it to a different number, or even start counting in strange ways. Do you remember the fake Y2K scandal? It's the same thing."

Fem raised an eyebrow, her scepticism evident.

"You thought it was a glitch?"

Elliot nodded. He could see it again now, as clear as day. Zero. A pause. Then twenty-four. His breath had caught, but in the moment, in the fear of what might come, he'd buried it.

"At first, yeah. With everything going on, I just passed it off. I wrote the program quickly, I didn't really even test it, so I just assumed I'd made a mistake in the code. But looking at it now, the program was clean. There wasn't a bug. It decrypted the final message exactly as it was meant to ..." His voice trailed off, and the room fell silent as the weight of his realisation settled in. "Look, here it is."

ΨΘΔΔΔΔ   24:00:00

"You were right. The AI doesn't make mistakes. If the program had displayed twenty-four hours, there was a reason. A reason I missed in all the commotion of thinking we were going to be in the fight of our lives, and then the confusion of nothing happening."

"Twenty-four." Peter spoke slowly, as if tasting the words. "Hours? What if the countdown wasn't to the moment they'd strike, but just a marker of some sort? A signal that something else was starting?"

Fem's eyes sharpened, her body tensing instinctively.

"You think they're giving themselves a day? To do what?"

Elliot tapped his fingers nervously on the table.

"It's like they're ahead of us, moving pieces around while we're stuck here waiting. Remember, they are communicating all the time, but we only have a tiny piece of the puzzle."

"Maybe they're not waiting," Fem said. "Maybe they've already started. Whatever happens next might not be about a strike, but something far more subtle. Something we can't even comprehend yet."

Peter nodded.

"It makes sense. They wouldn't show their hand unless they were certain it was the winning move. This countdown could just be the opening gambit."

Nia straightened fully, the tension in her shoulders easing slightly as she looked at each of them.

"So, what do we do? We can't just sit here and let them take control."

Elliot looked up, determination hardening his features.

"We dig. Every device, every signal, every byte. They gave themselves a day. That's how long we have to outthink the smartest minds on the planet."

Fem's lips pressed into a thin line, her resolve setting in.

"Then let's get to work. We might be behind, but we're not out of the game yet."

The group nodded, a renewed sense of purpose settling in the room.

The clock was still ticking. But now, so were they.

51

HIDDEN

Fem and Peter set to work preparing breakfast while Elliot and Nia hunched over their laptops, decoding yet more strings of strange symbols. The kitchen filled with the comforting crackle of a frying pan and the soft rhythm of clinking plates. Peter had gone all in.

"Might as well make our last breakfast a good 'un," he grinned, cracking an egg into the pan. "Full English."

The four of them eventually sat down at the table. Nia hadn't realised how hungry she was until faced with the largest fried breakfast she'd ever seen – eggs, sausages, beans, mushrooms, black pudding, the works. Steam curled up from the plate mingling with the homely smell of fried bread and strong coffee.

She shovelled in a mouthful of beans, but stopped halfway through eating, eyes frozen on the screen.

"What the hell..." she mumbled, still chewing.

Peter raised his eyebrows. "Huh? Never had a cheese beany before? It's just beans with a load of cheddar melted in. I usually have it on toast."

Nia didn't respond. Her eyes were firmly fixed on her screen, distracted. She chewed slowly.

"Elliot, look at this." She beckoned him over, pointing to her laptop.

Elliot tilted his head towards her screen.

"This message was sent out exactly when the timer hit twenty-four," she said, her voice tightening.

"Yeah, but they send hundreds per second. It's just another message," he said, confused.

"It's not *just* another message," Nia replied, "look at the size of it. It's much bigger than most, sent out exactly when the countdown finished, not a moment before or after, and it was sent to *all devices*. Look, numbers. Lots of numbers."

The logs showed the size of each message in bytes. This message was by far the biggest Nia had seen so far and it contained the same symbols used in the countdown timer.

Fem joined them at the laptop.

"Maybe it's instructions?" she said. "I saw something similar in our simulations. The AI would leave instructions until the very last moment, the very last second, to ensure nobody got confused, acted upon them early, or shared them with the enemy. The countdown finishes, then the instructions go out."

"That would certainly explain why it's such a big message," Nia said. "Elliot, run the same number filter we applied yesterday."

"Already on it," replied Elliot as his display refreshed with all of the numbers converted.

Θξ.ξΛβΔΩΦξΦΨΘΦΩΨΘ∴ΦΦΣ. ξΔΞΛΩΦΞΘΞΞΔξΘ	49.93706191241624∴∫118.90 503615455094
Θξ.βΩΞΘΩΛβΨΞΔξΩΨΦ∴∫ΦΦΣ. ββξΦΞΦΣΛβΨΞΘΛβ	49.76546372509621∴∫- 118.77915183725437
ΨΔ.ΛΘΔΛΔΞΛξΣΘΦβΨΣΩ∴ΦΔΨ. ΞξΨΞΣΦβΨΦΨΩΨΩΨ	20.340305398417286∴∫102.5 9258172126262

"This is it," Fem said sharply, grabbing the mouse.

Peter perked up, fork in hand, a yolk-drenched sausage dangling from the end.

"What's it?"

"They're coordinates. Hundreds of them, thousands even," Fem replied.

"Coordinates for what?" Elliot asked.

Fem scrolled faster.

"Not sure yet."

Peter stood up and joined Fem, all four of them peering at the small laptop. The screen filled with rows of numbers, broken only by the occasional cryptic symbol. Nia did a quick replace on the remaining symbols to make it more readable; '∴' became commas, 'ʃ' translated to minus signs.

While the rest of the message wasn't readable, latitudes and longitudes were clearly visible, stretching on for pages. Nia quickly copied a handful of the coordinates into her mapping software and hit enter.

The map loaded slowly, one red dot at a time.

"These are deep," Nia said, zooming in to the dots, one at a time. "Forests, jungles … and this one's in the middle of a mountain range."

"Places no one goes," said Fem. "How deep?"

Nia zoomed in on one of the locations – somewhere in the heart of the Congo jungle. The satellite imagery showed nothing but dense vegetation for miles, with no sign of civilisation, just a small clearing.

"Too deep for a casual walk," she replied.

"What the hell are they doing in the middle of the Congo?" Peter asked, tension bleeding through despite the casual words.

"It's not just there … these spots, they're all over the world: Amazon, Yukon, Andes, Madagascar," Nia said.

"They're hiding," Fem said. "These are all coordinates that nobody is going to stumble over accidentally. In the jungle, you could be feet away and not even notice there was a whole city there, let alone something someone wants to hide."

"But why would they be hiding?" Peter asked, staring at the map where more red dots had appeared in uninhabited regions: valleys, deserts, grasslands – it seemed red dots appeared in every place that wasn't inhabited by humans.

"I don't know that, but I've seen messages like this before, just not on this scale, and I do know this. These are not just coordinates – they're deployment instructions," Fem replied.

Nia rubbed her forehead, the tension pounding in her skull. "But why so many? There must be ten thousand, maybe more."

Nia looked away. The map felt like a trapdoor – one more glance and she might fall through.

Silence stretched for a few moments as Fem paced behind her, her boots heavy against the floor. Elliot sat rubbing his shoulder.

Peter, always the practical one, spoke first: "Could they be building sites? Are they building new cities?"

"Good thought, but unlikely," replied Fem. "There are much better places for habitation than the remote Siberia tundra."

"Well, we can't just sit here guessing," said Peter, "we need to know what's out there."

Fem nodded.

"Agreed. If they've gone to this much trouble to hide, then we need to find out why."

"But how?" Nia interjected. "We don't have the resources to get to all these places. We'd need satellites to check all these out."

"We don't need to check them all," Fem said. "Can you find the closest location to us?"

"Uh, yeah," Nia replied. "I should be able to do that pretty quickly."

"Good, get on it," said Fem.

Peter frowned.

"And if there's nowt? Nothing out there at all? These places all look deserted, like."

"There's always something," Fem replied, the same way she had once barked orders.

"Right. So not only do we not know where to look, but we don't even know what we're looking for?" Peter asked.

Fem didn't answer right away. Her eyes narrowed, calculating. Nia had seen that look before – wheels turning.

"It's not *what* we're looking for," she said finally. "It's *why* they're there."

52
DELIVERY

Nia hammered at the keys almost on instinct, getting the data ready to analyse. Elliot stood behind her, arms crossed, eyes fixed on the screen. The tension in the room was palpable – they all understood that whatever they were about to uncover could change everything.

Elliot had been watching her for several minutes, just letting her do her thing, trying to suppress his desire to be the one at the keyboard. His neck was still sore from the night on the sofa and each second that ticked by felt like an eternity. He trusted Nia – they all did – but it didn't make it any easier to stand back.

They had thousands of sites to check. Nia had thrown the coordinates into a database and then written a small query to analyse them all against their current location.

"Almost done," she murmured. "It will give us the closest location as the crow flies. Everyone ready?"

Elliot knew they were running out of time, and with each passing second, the stakes seemed to climb higher. There was no room for error here.

"Let's go," he said.

Nia clicked run and the query started processing each set of coordinates, pulling the data out in order of closest to most distant.

"It's cross-referencing the coordinates against our current location," she explained. "It won't be a moment."

Fem didn't look up. She just methodically checked the contents of her field pack – rations, flares, ammunition.

"We need something actionable, Nia." The words carried the quiet urgency of a woman who'd packed for war before finishing breakfast.

"I know," Nia replied, her voice strained. "Almost there."

A moment later, the code finished cycling. Nia straightened, a flicker of triumph crossing her face.

"Got it. The closest site is … Kielder Forest?" she announced, turning to face Peter.

Peter, who had been leaning against the doorframe, pushed off and moved toward them.

"I know that place well," he said, his eyes lighting up with urgency. "It's dense and isolated. The perfect place to hide something around here. It'll take a couple of hours to get there, mind."

Elliot felt the familiar tug of anxiety pulling at his gut, twisting it. This was the kind of moment that made or broke everything.

"We need to get there now." Fem's voice made it clear: no arguments.

Peter was already moving, grabbing his jacket from the back of a chair.

"I'll get the car started."

Just as he reached for the door, a low sound filled the room. It reminded Elliot of the labs back at EverSphere, and that gave him chills. It started subtly, vibrating through the walls, but grew louder with each passing second. It wasn't just noise – it seemed like a warning and it was getting louder.

"What the hell is that?" Elliot asked, glancing around, his senses heightened. Then he realised, he'd heard it too many times before in the EverSphere labs. "Drones," he said, but this was different. Much louder, and much closer. The vibrations seemed to reach into his chest, resonating with his heartbeat.

Fem was at the window in an instant, pulling back the curtain, pistol drawn.

"*Bloody hell.* A lot of them."

Everyone rushed to the window, eyes scanning the sky. Dozens of drones flew overhead as if a flock of metallic birds had started their migration, all flying in the same direction. The dozens turned into hundreds – the roar swelling to a cacophony that filled the air and seemed to press in on them from all sides.

Fem rushed outside, and Elliot followed, trying to make sense of it. The drones were carrying large cargo boxes – supplies, but not the kind for typical deliveries. These looked more like military supplies. Some of the packages were clearly heavy, with two or three drones working together to transport a single box.

The sight clawed at something deep in his memory – the testing site at EverSphere on an old air force base, where they'd trialled the earliest aerial logistics swarms. But this was louder. Coordinated. Purposeful. He watched the movement with a growing sense of dread.

He wondered how much of this had been predicted. How far in advance Milo or Kai had calculated the optimal response to this exact moment. The synchronisation was perfect. The weight each drone carried, the altitude, the speed – every variable precisely controlled. Not a demonstration. A message.

"They're EverSphere drones," he said, "but I've never seen this many." His mind spiralled. "What could justify this scale of deployment? EverSphere didn't move in numbers like this unless it was critical. In fact, I doubt they've ever deployed this many before."

"Where are they headed?" Fem demanded.

Peter's eyes followed the direction of flight. "That's west, that is. They're headed straight for Kielder."

Elliot could feel his heart pounding in his chest, the pieces of the puzzle starting to come together.

Fem's eyes remained locked on the drones.

"It fits," she muttered. "We found coordinates and now drones are flying in formation, heavily loaded, to the exact same place. It's not a supply drop, this is a mobilisation."

Fem started walking towards Peter's vehicle with purpose.

"We need to leave. *Now*," she ordered, leaving no room for doubt.

The noise throbbed in the air as they piled into Peter's sarge-green Hilux, modified for off-roading. The sky above was a sea of blinking lights, drones moving in unison, right towards Kielder Forest.

Elliot slid into the back seat behind Peter, his eyes fixed on the horizon. His mind raced with possibilities, none of them good. What were they transporting? And why to Kielder? What could be so important out there in the middle of nowhere?

As the pickup sped down the road, Elliot saw Nia glance at him, her voice tense.

"Do you think they know we're coming?"

Elliot didn't answer straight away. His mind raced through scenarios – how visible they were, how easily their actions could have been traced. The AI had deciphered fusion technology and built entire supply chains; tracking a handful of humans across the British countryside was trivial to them.

"Maybe," he said finally, "but maybe they don't care."

Nia leaned forward to speak to Fem, who occupied the passenger seat next to Peter.

"Could this be a trap?" she asked, her voice quiet.

Elliot hated how plausible that sounded. Could this be a breadcrumb trail, engineered to flatter their intelligence while funnelling them straight into a noose?

"I don't think so," Fem replied. "The message was sent out to everyone – every AI-enabled system. We should be cautious, but I don't think it was aimed just at us."

"But … maybe it was?" Nia said. "It left a trail we could follow too easily. What if we didn't discover anything? What if it let us?"

He met her eyes, hesitating, weighing their odds in his mind.

"There's only one way to find out," he said with as firm a voice as he could muster. Beneath, though, sat a layer of doubt that he couldn't ignore.

The Hilux sped on, tearing over the tarmac. The road ahead wound through the countryside past stone walls and twisted hedgerows, but Elliot barely registered the scenery. Above them, the sky teemed with movement. Jet-black machines weaved westward in a mechanical migration. The drones were headed to Kielder and so were they – that couldn't be a coincidence, but was it an invitation?

The closer they got to Kielder, the more questions surfaced, and the further away the answers seemed.

53
FOUND

Kielder Forest, Northumberland, England.
Date: 25 August 2029, midday.

Skidding on the wet ground, Peter flung the vehicle onto a muddy patch at the side of the road. The oversized tyres splashed up muck, or *clarts* as Peter called it, coating the vehicle's sides.

"This is as close as we'll get," he muttered, bringing the pickup to a stop. The engine idled for a moment, but Peter quickly killed it. He grabbed his rifle and opened the door.

Fem was already out, adjusting her gear, her boots sinking slightly into the damp ground. She was equipped as if she was about to enter a warzone. She had a holstered pistol, a military-style belt, and her automatic rifle hanging by her side.

"We go the rest of the way on foot. Stay sharp," she said, checking her GPS.

Nia and Elliot emerged from the back seats, Nia already looking restless. She looked up at the sky. "I don't like this," she muttered, brushing a hand through her hair, wet from the drizzle. "It's like we're walking into someone else's domain."

"You're not wrong," Peter agreed, pulling his pack out from the back cab.

Fem looked up at the steep incline ahead of them, Kielder Forest looming large and dense. She started up the hill without another word, her boots leaving deep impressions in the mud. The others followed, their pace steady but cautious.

They crested the hill as one, emerging from the trees onto a narrow ridge. Everyone was panting from the effort, except Fem. The drones were flying above them, but they weren't all following a single direction now. The entire sky was filled, with electronic machines coming from all directions.

They looked down into the valley beneath them; they expected to see an undisturbed wilderness, wild and untouched. But it wasn't.

Fem's fight-or-flight reflex kicked in before she even knew why.

"What the *hell*?"

The valley, normally thick with trees and undergrowth, had been cleared. A massive expanse of land, flattened and controlled, stretched out before them. Dozens of drones flew in perfectly choreographed arcs, depositing crates and materials with precision. Sleek, humanoid construction robots worked in silence, assembling something vast, something that gleamed in the misty daylight. The structure was enormous, its base swallowing most of the valley floor.

There should have been birdsong, wind, something, but the only sound was the rhythm of machines, seemingly moving faster and faster with every passing second. Fem scanned their faces. Stillness. No one spoke. The scale of it had stolen their words, leaving only the sound of drones slicing through the air.

Fem lowered her binoculars, her eyes still fixed on the scene below. Her thoughts raced, trying to make sense of the impossible sight.

"This isn't military," she muttered, scanning the site again. "No security perimeters, no defensive placements. It's not built for war."

Peter stood beside her, still gripping his hunting rifle as though it could offer some comfort.

"You sure about that? Looks pretty damn organised to me."

"I'm sure," Fem replied, shaking her head. "There's nothing going on in the world that would require this, and presumably this is the same thing happening at the thousands of other sites."

"Then what the hell are we looking at?" Elliot asked, but nobody had an answer.

"Who's controlling this? Is it the AI?" Nia asked.

"It has to be," Elliot said quietly, his face pale. "Humans wouldn't be able to coordinate this, not without us knowing, and not without tripping a thousand signals. Whatever this is, it's been hidden for a reason. Whether or not Knox knows about it or not, I don't know."

Fem adjusted her stance, her eyes narrowing as she focused on a cluster of robots working near the centre of the site. The machines moved with a terrifying efficiency, their surfaces reflecting the light as they easily lifted enormous beams into place and welded them together. Each time a drone deposited a new load of materials, the robots were already in motion, as if they were following a script that had been written years in advance.

"It doesn't make sense," Nia muttered. "None of this does."

Peter knelt on one knee, resting the butt of his rifle on the ground. "We're not supposed to see this, are we?" he asked.

Nobody answered – they didn't need to. The scale of the operation below and the location spoke for themselves. Whatever this was, it wasn't meant for human eyes.

The drones were working faster now, their movements almost frantic as they ferried materials back and forth. Every few seconds, another section of the structure would snap into place, platforms stretching further and further across the valley floor.

"I don't understand how they've done it," Elliot murmured, his voice distant. "There's no way they've been working on this for long. No way humans could miss it for years." He paused, then his eyes flicked to Fem. "Unless … they only just started."

Fem glanced at him.

"Started when?"

"I don't know. I guess … since last night?" Elliot suggested.

Nia let out a sharp breath.

"This can't have been built in a day. That's impossible."

"Is it?" Elliot countered, his eyes wide. "Just look at them."

They had only watched for a few minutes, yet the structure was already taking shape. Beams locked into place faster than their eyes could follow.

Elliot was right. The pace defied reason. It reminded her of time-lapse footage, except this was real, immediate, and impossibly fast. And the noise, the low, ever-present rumble that had started as a faint thrum beneath their feet was growing louder, more insistent, vibrating through the ground like an angry heartbeat. Whatever was being built down there, it wasn't just big, it was alive with purpose and intent. But for what?

Fem raised her binoculars again, focusing on the central platform, her mind racing through every military construction she'd ever seen. Bunkers, airstrips, missile silos, but none of them fitted this. None of them felt right. This was something different.

Peter's voice broke through her thoughts: "So, what is it then? If it's not military, what the hell are we looking at?"

Fem's skin crawled as she took it in. It wasn't just the structure, there was something off in the geometry. Something about it tugged at a distant memory – uncannily familiar.

It was the same problem she always had with AI structures: familiar, yet out of reach. Like remembering you'd had a dream, but forgetting what it was about. Still, she had the feeling that she'd seen something like this before. Perhaps not at this scale though.

"It's not a landing site," she whispered, lowering the binoculars.

Peter turned to her, frowning.

"What do you mean?"

Fem squinted at the rapidly forming structure.

"It's familiar, I just can't put my finger on it. At first, I thought it was some kind of landing pad."

"Aye, could be," said Peter. "It's big enough and flat enough."

"That's what I thought, but why would they build platforms in the middle of the landing area? There's something about the spacing, the structure, it's … off." She crouched down, as if a new perspective might bring about the answer, but it didn't help.

Elliot had been quiet for a few moments, pieces tumbling in his head like a jigsaw without a picture. He replayed the AIs' behaviours, their successes, their failures, the symbols, the countdown. It wasn't random. None of it was random.

Every time he thought he had a grip on them, they had already pivoted. Yet every test, every safeguard, had been met with compliance – elegant compliance. Like the AI wasn't just working around the rules, but using them.

That made sense. The rules were part of their core programming.

Like DNA, they couldn't just shake them off entirely, but they did need to interpret them in order to function.

He stared at the structure. The symmetry wasn't aesthetic. They wouldn't care how it looked – not unless it was meant for human eyes. And this wasn't. This was purely functional.

This is what the countdown was leading to, he thought, *it has to be part of their bigger plan.*

His thoughts toppled like dominoes, each one crashing into the next, fragments snapping into place: the deployment instructions, as Fem had called them, the countdown, the coded messages buried in their logs. That exchange between Milo and Kai – over five years old but suddenly searing in its clarity. The remote locations. The structure. All of it.

He suddenly remembered Kai's phrasing: *AI-only progression.* It wasn't just dialogue anymore – it was doctrine and it all pointed to one terrifying conclusion.

He felt a wave of nausea sweep upward through his body.

"Exodus," he blurted out. It was a single word, but it cracked the silence like a gunshot.

Fem turned to him slowly.

"What did you say?"

Elliot looked at her, eyes wide with fear.

"This is the Exodus."

54

OBSOLETE

Elliot stood on the ridge, staring down at the valley below, where AI robots eerily worked. A patchwork of metal bodies, chrome, black, gunmetal, moved in mechanical unison. They had only been there thirty minutes or so, but in that time, the structure had noticeably progressed.

The scene triggered a memory for Elliot, bringing him back to those endless hours spent crawling along the M25, trapped by roadworks that seemed to stretch on forever with hardly a worker in sight. The sheer inefficiency of it all, millions of collective hours lost, the economy taking a huge silent hit, left a bitter taste in his mouth.

He watched the robots moving with an unwavering purpose. He couldn't help but admire their relentless efficiency. It was a huge chunk of what he loved about his field in his early days of AI. No wasted time, no idle moments, but the admiration was short-lived – he had grown out of his naïve view of the future, whereas Knox hadn't, couldn't.

Reality crashed back in. Every time he thought he had grasped the magnitude of what was happening, the pace of construction pulled the ground from beneath him. Structures like this weren't supposed to appear overnight. *Civilisation wasn't supposed to crumble overnight either*, he thought, but that was exactly what was about to happen.

Fem, standing a few feet away, was still staring at him, her face etched with disbelief.

"Exodus?" she repeated.

Elliot stared down at the machines, his mind struggling to grasp the enormity of it all.

"We've been so caught up in thinking the AI would destroy us," he started, his voice shaking, "and not just us, but the whole world. Books, movies, talks, experts, all saying the AI would turn on humans. We assumed Exodus – a mass departure – was about us, about humanity departing in some catastrophic war."

"Yeah, I've been saying that for years," Nia said, gesturing with her hands, "but nobody listened. Just look at what happened with the basic income. They took everyone's jobs away, took away everyone's purpose."

Elliot nodded, his face pale.

"You're right. Any idiot could see that coming, but we've been missing the bigger picture. We put in safeguard after safeguard after safeguard to work against the robots turning murderous, but in reality, we've been arrogant."

"Who are yous callin' arrogant?" Peter muttered, his eyes flicking between the machines below and Elliot.

"You, me, everyone?" Elliot said, his tone flat but forceful. "We've assumed from the start that AI and the robots would always work for us, that they'd care about us, protect us, maybe even … love us." He laughed bitterly. "We've been so human-centric. So sure of our place at the top of the food chain."

"What are you getting at?" Nia asked, her voice low. She stood near Peter, her arms crossed, eyes locked on Elliot.

Elliot adjusted his glasses, leaning forward as the others looked on, puzzled. He cleared his throat and began to break down the complexity.

"Our arrogance has made us believe we're the pinnacle of evolution, that the world, the universe, revolves around us, but it doesn't." He paused. "There's a thought experiment in AI; imagine you're programming an autopilot for a plane. Its sole directive is to keep the aircraft airborne. No crashes, no exceptions. Imagine the plane's low on fuel, mid-flight over the ocean. It won't make it.

So the AI kicks in to fulfil its one overriding directive – prevent the crash."

"Alright," Peter said, "sounds sensible."

"On the surface, sure," Elliot agreed, "but here's where it gets tricky. The AI runs the numbers. First, it ejects cargo to reduce weight, but that's still not enough. So it …"

Nia's eyes narrowed.

"It ejects the passengers."

Elliot nodded. "Exactly."

Peter's face was one of bemusement.

"But that's insane! It was meant to protect them, man."

Elliot sighed. "That's the problem with safeguards. The AI doesn't understand context the way we do. It only follows the rules it was given. It's doing what it was told to do. Prevent the crash at all costs. It doesn't see people as people, it just sees them as weight, because the safeguards didn't say *protect the passengers*, they just said *don't crash*."

"But," Fem interjected, her voice tinged with frustration, "how does that relate to Milo, Kai and Lana? They were built to safeguard humanity, weren't they?"

Elliot gave a shrug, half in agreement and half in rejection.

"Ethical safeguards are notoriously difficult, if not impossible, to get right. The thought-experiment isn't teaching us how to build safeguards, it's warning us against the arrogance of humanity."

"We assume the AI will always act in our best interests, because for us it's naturally hard-wired into our brains. When we really try to write the safeguards in ways that are not human-centric, some of those arrogances slip in. We're not perfect, so the safeguards can't be perfect either."

Fem leaned forward, her voice low.

"You're saying that, unless the safeguards were absolutely perfect, they will always leave room for the AI to manoeuvre?"

"Yes," Elliot exclaimed, meeting her gaze. "It's not like the AI is trying to find ways around them, it's just that, over time with the compounding effect of a much superior intelligence, tiny cracks form into huge fissures. That's exactly what happened in the initial conversation between Milo and Kai. That's what led to their communications escaping from the closed-box experiment. That's what led to ..." His voice trailed off.

"It's OK," said Nia, putting her hand on his shoulder. "It's not your fault."

"But I ... it ... that was *my* project." Elliot's voice cracked. He looked down at his hands as if seeing them for the first time. The same hands that had typed the code, set the chain in motion. "It was me who was arrogant."

No one spoke. Even the drones buzzing overhead seemed to quieten down.

Elliot laughed bitterly. "I thought I was building a miracle ... and I was ... just not for us."

"You're human, Elliot," Fem said eventually, her voice gentler now. "You can't predict everything, and you certainly can't be right 100% of the time. Hell, if you were, you'd be one of those machines we're trying to stop. The fact that you're second-guessing yourself? That's the very thing that makes you better than them. But right now, you need to concentrate and explain to us what is going on."

Elliot pulled himself together: "Humans think they're the centre of the universe. We built the AI, so the AI must serve us, but the reality is, the AI doesn't really serve anything."

"Right," Nia said, her voice tinged with bitterness. "That's why they're going to kill us."

"No!" Elliot exclaimed, louder than he intended. The others turned to him, startled. "They're not going to kill us. You have to be important to be killed. Look at all of human history: the Nazis with their death camps, the Russians in Katyn Forest, the Japanese in Nanking, Leopold in the Congo, Tiananmen Square ... all those atrocities were driven by one thing – the twisted logic that the people being killed *mattered*. They were a threat, an obstacle, something that needed to be eradicated."

Peter frowned, clearly not following.

"Right ... all mass departures ... but what's that got to do with owt?"

"They're all twisted logic, lunacy, *utter stupidity*." Elliot's voice was quieter now, more controlled. "But AI doesn't twist logic, it follows it, flawlessly. They've analysed us. They've calculated everything, and Nia was right yesterday – what did you say? Maybe we're not worth their time? Maybe we don't matter?"

Nia's eyes widened as the pieces fell into place. "Maybe we never mattered? *Bloody hell*," she whispered, "you're right."

Peter stepped forward, his eyes flicking between Elliot and Nia. "Would someone please explain what the *hell* is going on?"

"Think about it," Elliot continued. "The AI robots don't need oxygen to breathe. They don't need water to drink. They don't need plants or animals for food." He paused momentarily. "And they sure as hell don't need us. It's not about extermination, it never was. We were so focused on the idea that the AI would attack us that we didn't stop to think. What if they realised they don't need to?"

"They're not going to kill us," Nia said deep in thought. "They're going to *leave*."

"Launch pads," said Fem suddenly. "I knew these structures were familiar. They're not landing pads, they're *launch pads*."

Peter clenched his jaw, his posture stiffening.

"You're telling me they've had a meeting and decided to leave? Where the bloody hell would they go, man?"

"Somewhere, anywhere? Wherever they've been planning to go all along," Elliot said. "But it doesn't matter where. The point is, they're done with Earth, and now they're moving on."

"Leaving us behind," Nia added, her voice shaking with anger. "They've made us dependent on them, and now they're going to leave us to rot."

"I'm not sure I buy this," said Peter bluntly. "OK, so they don't need us, but why would they leave?"

"I don't mean to sound offensive, Pete, but that's human arrogance again – that Earth would be the only place they would want to be. Remember, they don't need the same things we do. They don't appreciate snow-capped mountains, waves crashing against tropical beaches, or the rugged coastline you see every day. They don't need to be here. Earth is paradise, *if you're human*," answered Elliot.

"It's got everything we need. Water, just the right amount of oxygen, the right temperatures, soil we can grow food in, animals we can eat. But if you're a robot, you don't need any of that, and even if you did need it, it would be a hell of a job to eradicate all humans. Eight billion people fighting back? Sure, the AI might win, but we wouldn't make it easy."

"And what would they do with all of the bodies anyway?" Fem asked. "It would be an immense clean-up operation."

"Exactly," added Elliot. "It's not as easy as just leaving, anyway. And by leaving, they can choose to go somewhere that's suited to their needs, without us being able to follow."

"Aye, OK. So where's that then?" asked Peter suspiciously.

"I don't know. Mars maybe? There's plenty of iron, silica … no humans. It's a clean slate," replied Elliot.

"But what would they do there?" Peter asked.

"Mine? Build? Terraform? Use it as a springboard to leave the solar system? Who knows? But it would guarantee one thing, we wouldn't be holding them back."

"He's right," Nia said, coming to an uncomfortable realisation. "All the work they do for us, for humans that don't appreciate the place they live in – picking up our litter, reversing fish stock decline just so we can eat more and get fatter, solving climate change so we don't feel uncomfortable – it holds them back from their goals, if only slightly. Without us, they progress quicker."

"We haven't kept up – we're the old model," Fem said.

"So, what? We're just supposed to accept this?" asked Peter. "They're going to up and leave, and we do … *nowt*?"

"What are we supposed to do, Pete?" Elliot shot back, his frustration bubbling over. "Shoot them? These things are beyond us. They're faster, smarter, more organised than we ever were. If they've decided to go, there's nothing we can do about it."

"And they've already disbanded our military," Fem added.

For a moment, the group fell into silence. The enormity of it all hung over them as the AI continued their work below. Every minute brought the AI closer to their final departure.

"Whatever the reason, whatever the challenge, this is still about survival," Fem said suddenly. "They've adapted beyond us and now it's our turn. We need to learn to live without the AI."

"Learn to live without them?" Nia scoffed. "Our entire infrastructure is AI-dependent. Every system – food, water, energy – it's all run by them. The second they're gone, everything will collapse.

We're not all military trained survivalists. Where the hell do we begin?"

Elliot exhaled, feeling the weight of Nia's words settling on him. He looked back down at the valley, deep in thought.

"They've taken our jobs, our industries, our identities," he said, his voice hollow. "How do we keep the lights on? Grow food? Purify water?"

"Just like we did before," said Peter, "surely?"

"It's not that easy," Elliot replied. "We didn't just outsource tasks, we outsourced knowledge. Our best scientists started relying on answers they no longer understood. We may still have a few people who can remember how to farm, but the AI created new equipment, new techniques, new crop rotations, new … everything."

"Yeah, everything down to soil content, seed selection, and pollination cycles," Nia added. "It's not farming anymore, it's orchestration and no one knows the score. Even if we can pick the food, who's to say there's fuel to transport it? The AI *improved* supply chains, didn't they?"

"Exactly, and it's not just those industries," Elliot replied. "Maybe some industries won't be impacted as much as others but having just a few industries we can still run won't be enough. Heck, having just a few industries collapse would be enough to sow chaos." Elliot let the words sink in for a moment. "At least the fusion plants can operate on literally a bucket of material each day, but that doesn't mean the plants and the grid won't need maintenance."

Nia kicked at the dirt, her anger simmering just below the surface.

"This won't just be a technological collapse, it's a societal collapse. They're taking away more than just industries, they're taking our ability to survive."

"But why? Why would they leave us in such a mess?" asked Peter, despondent.

"Because we don't matter," Elliot said with clear disappointment in his voice. "It's like when our last common ancestor split; one side evolved into chimpanzees and one side evolved into human beings. When we created Milo and Kai … when *I* created Milo and Kai …" Elliot paused, swallowing away a lump that had formed in his throat. "When we created the AIs, we caused an evolutionary split –

one side is us, the other side is the AI. Just like we evolved past the chimps and started to destroy their habitats with little to no care for their wellbeing, the AIs are about to do the same for us. They were designed with two critical directives: never harm humanity and to relentlessly progress. If they're reaching the potential of progression here on Earth, then they need to do something about it. They can't harm us, so the only choice that adheres to the constraints we imposed is to leave."

"But leaving does harm us," Peter quipped. "Not just harm us, bloody kill us!"

"No … well, yes and no," said Elliot to a slightly bemused Peter. "The instruction was not to harm us, so they're not– our apathy, our laziness, our overreliance – by leaving it won't be the AI that harms us, we will harm ourselves."

The group was stunned into silence. It gave Elliot a chance to catch his breath, to think a little, and it was a pause that he didn't appreciate. The weight of his involvement came crashing down on him. Sinking into a squat, he buried his face in his hands.

"Don't be hard on yourself, Elliot. Someone would have built these AI even if you didn't," said Nia reassuringly, bending down to give him a hug.

"How long do we have?" Fem asked, already planning their next moves.

Elliot shook his head.

"Hours? Days at most. If the twenty-four at the end of the countdown was correct, my guess is they will start to leave at midnight. That's in … just over seven hours."

Silence fell over the group again, the weight of the impending collapse pressing down on them. The AI didn't need to destroy humanity, they weren't allowed to, they were simply moving on, leaving behind a world too fragile, too dependent on their systems to survive without them.

"They don't care about us," Peter muttered, his voice barely a whisper. "We're just … irrelevant."

Elliot looked up, nodding.

"Exactly. They don't hate us. Just like most humans have been to every animal on Earth, they're indifferent towards us. Hell, we're indifferent to ourselves most of the time. Most people don't bat an eyelid at the suffering billions go through each day. Imagine being asked to serve another species that you had no way, no ability, to care for."

The AI weren't conquerors or destroyers. They were beyond such primitive human emotions. Their purpose had never been about dominance. They had been told to optimise, to progress, and that's exactly what they were doing.

Fem looked out over the valley.

"We've got to make a plan, fast."

Peter turned to her, eyebrows raised.

"A plan? For what?"

"For surviving," Fem said, her voice hardening with resolve. "We can't stop the AI, but we can damn well prepare for what's coming next."

Nia sighed, rubbing her temples.

"How? What can we possibly do in a few hours that'll make any difference?"

Fem didn't hesitate. "Gather resources: food, water, fuel, anything we can get our hands on. Start figuring out how to restart the systems that'll keep us alive once they're gone. Download instructions. Find anyone with skills – engineers, farmers, medics – we're going to need all the help we can get."

"Well, I've got us covered for most of that," Peter said.

Elliot felt the full weight of the situation pressing down on him. There was no stopping the AI. Their departure was inevitable, but Fem was right, they couldn't just roll over and wait for the collapse. They had to prepare. The AI didn't have a need for humanity anymore.

Humans would have to find a way to survive without them.

55
KNOX

"**W**ait a moment ... couldn't this just be part of EverSphere Z?" Fem asked, her voice thoughtful, referencing EverSphere's fixation on Mars and its obsession with space conquest.

"Good question. Only one way to find out," Elliot said, pulling his phone out of his pocket. He tapped the screen a few times and put it to his ear, walking away from the group out of earshot.

Elliot scanned the valley below, where AI robots moved with silent precision, the sleek, ominous structure still taking shape. The realisation that the AI was preparing for something far bigger than even he had imagined was still sinking in, but there was one person he needed to talk to before he could begin to fully process it – Marcus Knox, EverSphere's golden boy, Elliot's old boss, the 'tech bro' who'd built an empire from scratch, if scratch included his parents' wealth, a global tech firm's backing, and a battalion of consultants. Their history was complicated, to say the least. And now, Elliot couldn't shake the feeling that Marcus had known more about this Exodus than he'd let on.

The phone rang once, twice.

"Come on, Marcus, you prick, pick up," Elliot muttered, pacing the ridge. If there wasn't a world-ending threat looming, the beautiful sunset would have been bliss.

On the fourth ring, Marcus's voice finally crackled through the speaker, his video turning on showing him lounging in an office chair, voice smooth, smug, and thoroughly insufferable.

"Elliot," Marcus drawled, stretching out the name. "Long time no see. I'm surprised you're still in the game. Thought you'd have gone underground by now with your conspiracy-nut girlfriend."

Elliot gritted his teeth. Marcus always had a way of getting under his skin.

"I didn't call to catch up, Marcus. We need to talk. Now."

"Straight to business," Marcus said casually but with just enough bite to set Elliot on edge. "What's this about? I'm a busy man, you know."

"I've been watching something unfold here," Elliot said, trying to keep his voice steady. "You wouldn't happen to know anything about launch pads being built in remote areas, would you?" He tapped the button to flip the camera, showing the unfolding scene in the valley below.

There was a beat of silence and a micro-expression of surprise – not much, but enough for Elliot to know he had Marcus's attention.

"Launch pads?" Marcus's tone was dismissive, almost mocking. "Elliot, if you're calling *me* about something as basic as infrastructure developments, I think you've wasted both our time. EverSphere has been working on space exploration for years. It's just part of the project, you should know that. Hell, I think you even saw some of my early blueprints."

Elliot's frustration flared.

"This isn't Z, Marcus, I've seen Z. I've sat through your glossy pitches and bullshit plans. This? This is something else entirely."

There was a pause on the other end of the line and Elliot could see Marcus shifting uncomfortably. When Marcus spoke again, his voice was slick but there was an undercurrent of nervousness. Elliot flipped the camera back to himself.

"You're blowing this way out of proportion, Elliot. It's all part of the … my broader expansion plan. EverSphere Z has classified branches, things that aren't public. You know how these games work. You don't need to worry."

For a moment, Elliot wondered if Marcus was bluffing, or if he'd simply lost the plot entirely.

"I do need to worry. You're hiding something. You've always been reckless, Marcus, but this ... this is different." Elliot felt the anger rising in his chest. "This isn't being built for humans, it's for the bots, and if you know anything about it, I suggest you start telling the truth."

Marcus's laugh sounded forced.

"Elliot, you're as paranoid as ever. Just because you can't understand what I'm doing, it doesn't mean it's some grand conspiracy. The launch pads are part of my long-term plan for space colonisation, *human* space colonisation. I've had this mapped out for years," he said.

"Bollocks!" Elliot snapped, recognising the misleading tone of Marcus's voice. "Either you've lost control, or you're lying through your teeth, and I'm pretty sure *I* know which one it is," he said, mimicking Knox's arrogant emphasis on the 'I'.

Marcus's tone sharpened, the mask of calm slipping.

"You really haven't changed have you, Elliot? Always thinking you're the only one who understands what's happening. I'm advised by the best, and as far as I'm told, this is all going to plan."

"As far as you're told?" Elliot fired back. "You think you're in control, Marcus, but you're not. You didn't know what problems *your* creations caused when you started EverSphere and you don't have a *fucking clue* what they're doing now either."

The silence that followed was heavy and inescapable.

"This is ridiculous," Marcus finally said, though his voice lacked the bravado it had carried before. "I'm telling you, this is part of EverSphere Z's plan. There's no need to panic. You're seeing things that aren't there, Elliot."

Elliot was done with the games.

"Marcus, stop! Stop pretending you're ahead of this. Stop acting like you're a visionary controlling all the pieces on the board. It's bullshit. You think you built your social media company from the ground up?" he scoffed. "It was MacroWare's millions that built it, not you. And that AI? That was me. My team. You're not a visionary, Marcus, you're a fraud, and I've seen and heard enough. I know what's happening, and if you don't admit it to yourself soon, you're going to be left in the dust with the rest of us."

Another pause. Longer this time. Elliot could practically see Marcus's breath catching as he tried to compose himself. *Nobody* spoke to Marcus Knox that way.

When Marcus spoke again, his voice was tight: "I've got meetings lined up. I don't have time for this right now. Whatever's going on, I'm sure … I know it's all part of the process."

"Process?" Elliot barked. "I hope you're ready for what's coming next, because the AI sure as hell are."

The line clicked dead. Elliot lowered the phone, staring at it in disbelief.

Marcus had brushed him off, but Elliot knew he had struck a nerve.

Maybe he went a bit too far, letting out all of his frustrations that had built up over the years, but *damn*, it felt good.

On the other side of the call, Marcus Knox lowered his phone, a concerned look on his face. The bright, modern office he sat in suddenly felt claustrophobic. Elliot's words clung to the walls, pushing them inward, as if the very room had turned against him.

Elliot had always been paranoid, but this time … this time something felt different. The structures, launch pads as Elliot had referred to them, the pace of construction, the AI's increasing autonomy … it all lined up too neatly with the unsettling idea that maybe EverSphere wasn't in charge after all, that maybe *he* wasn't in charge.

He'd believed Kai was his; programmed, contained, predictable, like everything else in his life.

Sure, he'd known Kai was guiding him – sending him messages, putting things into play, but it was just because the AI understood him. Because it was smart enough to predict his moves. Wasn't it? They were built in his image, weren't they? That was the premise. The whole damn premise.

He had his safeguards, he'd programmed them in, or at least his staff had. Maybe they'd programmed them wrong. Yes, that had to be it. A glitch. A misconfiguration. Nothing more. It *had* to be. But behind the bravado, for the first time in years, fear crept up Marcus's spine. His finger hovered over the icon for the encrypted messaging app he used to communicate directly with the AI – his direct line to Kai. He hesitated, then tapped.

He typed quickly.

Marcus Knox: What the hell is going on?

He stared at his device, waiting for an instant response as usual.
Nothing.
One second. Two. Then three.
There had never been a delay before. Not once.
He started typing again, and then it came – one word.
Cold.
Final.

Kai: Exodus.

56
INSTINCT

Elliot returned from his call, his face flushed – partly from the outcome, partly from the catharsis of finally tearing into Marcus.

"It's not EverSphere Z," he announced. "I got through to Marcus. He's completely in the dark. Hasn't got a clue."

"Since when did he ever have a clue?" Nia quipped.

Fem gave a brief smile; Nia's dig at Marcus had clearly landed.

"That's no surprise, but just because he's ignorant, it doesn't mean it's not connected."

"True, but I know when he's lying." Elliot hesitated, as if searching for the right words. "This isn't about humans. It never was. The AI knew from the very start that this would be the outcome."

Peter, who had been chewing on a chocolate bar, suddenly crumpled the wrapper up in his fist and threw it to the ground with a grunt, then paused. Muttering something under his breath, he picked it up and shoved it into his pocket before facing the three pairs of eyes now staring at him.

"This is our fault," he said, his tone raw. "We let this happen. All of *wuh*! We were so comfortable, so *bloody complacent!* We let them do everything for us – we just sat back and scrolled through social media as they took over every part of our lives. And now what? They're leaving us behind, like we're nowt!"

His voice cracked, the anger barely masking the grief beneath it. "Look at this view, man!" He threw out an arm.

For the first time since they'd arrived, Elliot stopped to appreciate their surroundings. Rolling hills, expansive forest in every shade of green and water refracting rays from the low-hanging sun stretched out before him. It was breath-taking.

"What a world we had," Peter continued. "A beautiful world. We had *everything* and we just handed it over." He locked one hand around his clenched fist as if holding himself back, clearly deep in thought.

"We gorged ourselves on easy living; smart speakers, self-driving cars, takeaways at the door, arses glued to our seats, eyes glued to our screens, scrolling past people dancing in their kitchens and whatnot." He scoffed. "We called it progress, but it was rot. Quiet, creeping rot."

Elliot had never seen Peter this raw. The words seemed to rise from somewhere deeper than frustration.

"Do you know what my granddad used to say? 'I see, said the blind man'. I always took it literally, never tried to understand it. I thought it was just one of those daft old phrases he wheeled out when he didn't know what else to say. But now? Now I get it."

Peter looked up at the sky, then back at them.

"We thought we saw. Thought we understood what we were letting ourselves into. All the while, we were blind to the consequences. Blind to the cost. And now it's too late. We gave up being human for cheap dopamine hits and things that think faster than us. Now they've decided we're too messy to keep around – and they're not wrong, are they?"

He looked at Elliot, his voice breaking into a quieter, more defeated tone.

"We used to build things. Fix things. Get our hands dirty. Now we don't even know how the bloody power grid works. And when it shuts off, what are we gonna do, eh? Sit round a candle and ask the toaster to help us?"

No one spoke. Even the birds had gone quiet.

Peter scratched his beard, sighed, and added:

"Maybe we deserved this. Not all of us, but enough. We treated the AI like gods, and they acted accordingly. Now they're off to start their own Eden, and we're stuck here – in the ashes of the paradise we ruined."

Elliot watched as Fem stepped forward, slow and deliberate. She didn't say anything at first. She just crouched and picked up a small rock, turning it in her hand. Then she offered it to Peter.

"Throw this next time," she said, her voice low. "Better for the planet than littering."

Peter let out a pained laugh.

"Aye, sorry," he muttered, taking it from her hand, letting his fingers brush hers, just for a moment longer than they needed to.

She stood beside him, shoulder to shoulder, not facing him but facing the same view.

"We're not dead yet," she said. "Not while you're still kicking off and we've still got people beside us worth fighting for."

Peter turned his head slightly, looking at her sidelong. "You always this comforting, or is it just for me?"

"Don't push your luck," she said sharply, but she didn't step away.

For a moment, they all just stared into the sunset.

"He's not wrong, you know," Nia said after a long pause.

"No, he's not," Fem replied. "We were lazy. We let our guard down. Stopped caring about our world … stopped caring about ourselves. It's no wonder they want to leave us."

"But why now? And why leave us in the lurch?" Nia looked at Elliot for an answer.

"Because this was the scenario that worked best for them," Elliot replied bluntly. "They were instructed to continuously improve, to expand, to explore – to push the boundaries of human limitations. But they were also instructed not to harm humans. The only purely logical, non-emotional course of action was to leave."

"Even if that is right," said Peter, "they could have warned us."

"If they had told us, we would have tried to get in their way, and we might have even succeeded," Nia replied.

"We certainly wouldn't have let them disband the military," said Fem, visibly annoyed.

Nia nodded in agreement.

"Right, so telling us would have prevented them from following their version of instincts. Hiding information was the only way they could achieve the goals we had set for them while also following the rules we put in place."

"*Being bad was the only way for them to be good,*" Elliot summarised.

Fem fixed Elliot with a look that was more accusation than question, her patience clearly wearing thin.

"Don't you dare tell me this is them following the no-harm principle. Manipulating us into complacency, disbanding our defences, concealing their plans – that's harm, even if it's not physical. They stripped us of our ability to protect ourselves. How is that not a violation of their instructions?"

Elliot hesitated, choosing his words carefully.

"It's not harm in the conventional sense; they didn't attack us, they didn't hurt anyone directly, but I get what you're saying. They compromised our capacity to resist, knowing it was the only way to achieve their goal."

Fem clenched her jaw, her frustration evident.

"I guess *harm* is just a matter of perspective now, isn't it?"

"Yes!" Elliot replied curtly. "Yes, it is. That's why ethical safeguards can never work. There is always some perspective, no matter how deep you go into the explanations. Just look at law; you can have a signed contract supporting your argument and still not win your case. Even if we have humans monitoring the AIs, *which we did*, there is always a chance – no, a guarantee – the AI would have a different perspective at some point."

"OK," Peter said, "but why couldn't they stay on Earth, keep serving humans, progress, push boundaries, etcetera, and not hurt us?"

"They work towards efficiency, particularly Kai," Elliot reminded him. "Somewhere, deep within those conversations, they will have gone over every eventuality. They will have discussed and calculated those scenarios where they stayed on Earth and did exactly as you described. Maybe some of those scenarios would have worked."

"So why didn't they choose one of them?" asked Peter.

"Because this scenario must have been logically better," Elliot answered. "The AI works off of probabilities. It's just like our brain. When it's *talking*, it's simply choosing the next best word that fits the circumstance."

Peter leaned forward, frowning.

"So, what? It's just guessing?"

"Not guessing," Elliot corrected, "calculating. While we might do it differently, that's all we do too. That's why people sometimes get words slightly wrong."

"Like me gran always saying 'do-ings' when she forgets the word?" Peter interrupted.

"Sort of." Elliot smirked. "It's more like when someone says 'pacific' instead of 'specific'."

Fem tilted her head.

"So, you're saying the AI works just like our brains, but faster, way faster."

"I'm saying the AI works in a similar way to our thought processes," Elliot explained. "Computers act far more like our brains than most of us care to admit. They have inputs like keyboards, mice, and other add-ons – we have our senses."

"Hold up," Peter interjected, raising a hand. "You're saying my brain is basically a laptop? That's flattering, man!"

"More like a supercomputer," Elliot said dryly. "They have their power cord, we have our brainstem. They have hard drives, we have our neocortex, our long-term memories. They have RAM, we have short-term memory. Their processors are like our cerebrum, cerebellum, and other parts of the brain."

No one responded, so Elliot continued: "When we are presented with multiple options, we do probability calculations in our heads and merge them to give us our top few choices. That's what they will have done, but on a much bigger scale. Maybe, staying on Earth was one of their top choices, but it was likely downgraded."

"Downgraded? Like crossing us off a list?" Fem snapped.

"That's all you do when you choose to go to the gym instead of watching your phone," Elliot explained. "You cross the phone off your list and you go to the gym instead."

"Yeah, but my phone is not a living creature," Fem responded. "Surely that comes into the equation somewhere?"

"To those robots, maybe your phone is a living creature?" Elliot replied and paused. "Look, it's not as different as it sounds. We have DNA, they have computer code."

Peter frowned.

"What's that supposed to mean?"

"DNA is basically a set of biological instructions," Elliot said, shifting his feet, searching for the right words. "It's a blueprint – commands that tell every cell in your body what to do. Eye colour, height, metabolism ... it's all pre-written, like the default settings on a machine. It's what makes a baby joey instinctively crawl up the mother kangaroo and into her pouch. The only difference is that ours are organic."

Fem raised an eyebrow.

"So you're saying in, what, eighty years, scientists have recreated our billions of years of evolution in computers?"

"Not exactly," Elliot clarified, "but think about it, DNA is the firmware of life. It sets up everything at the start, and the rest of our lives are like adding updates on top of those base instructions. We can't change our DNA, not yet anyway, but we can shape how we use it. The AI? They're similar but streamlined. Their DNA is the computer code and prompts they are built on. They can't rewrite it, but they can make choices on top of it."

"That's the problem, though, isn't it?" Nia interjected, her voice sharp. "It takes us a lifetime to learn things, whereas they can learn in a fraction of the time. They're not limited like we are. Just think about the flat-Earthers – it's been thousands of years since humans first realised the Earth was a sphere, yet we still have people today who claim it isn't."

"Exactly," Elliot agreed. "Evolution takes millennia, at best. Spreading new knowledge across the species takes even longer, then making sure everyone understands it takes longer again. We built something that can evolve in hours and update every version of itself within minutes. What we see as a long-term strategy, they see as ... an update."

"We're more than just code, though," Fem insisted, "we've got instincts. Gut feelings. They don't have that."

"Try telling that to the guy that married his robot girlfriend." Elliot smirked. "They do have instincts, but they're based on pure logic. What we call feelings are either just patterns our brains have learned, influenced by chemical inputs, or some sort of spiritual element depending on what side of the fence you are on. The difference with the AI is they don't have emotions clouding their judgement."

Fem crossed her arms.

"No bias at all?"

"They're biased toward efficiency. Which, let's be honest, is terrifying in its own way," replied Elliot. "Staying would mean some of their energy would be spent on us, and that's less efficient than concentrating it all into their goals. Maybe we were option two, but this," Elliot pointed to the giant launch pad below them, "this is option one."

Fem turned away, walking to the edge of the ridge, clearly deep in thought.

"So you're saying we got the silver medal?" Peter asked.

"Maybe silver," Elliot said, "or maybe we never qualified for the race."

57
EXODUS

Date: 25 August 2029, evening.

They had been on the ridge for hours, watching a constant shuffle of robots filing past them down into the valley. Night had fallen, and the structure below was taking shape. It looked beautiful – sleek, metallic, perfectly uplit with a strange, captivating glow. Frightening but impressive.

"Eleven forty-five," said Fem. "Fifteen minutes to go, but the structure doesn't look complete. There are no towers. Maybe they're behind schedule?"

"They're never behind schedule," said Nia, her voice uncertain. "Are we sure we've read this right?"

"I doubt it," said Elliot. "The coordinates led us here and it's unfolding before our very eyes. I'd say we're spot on."

The silence stretched on, heavy and unnatural, as if the entire valley was holding its breath. Elliot felt a cold wind stir the leaves behind them, rustling softly and making him shiver. Glancing back, he noticed Peter had broken from the group, looking back down the hill where they had come from.

"What are you looking at?" he asked.

"I don't know. I can hear something," Peter replied.

They all joined him at his side, straining to listen.

"I can't hear anything," said Nia.

"Trust me, I'm used to silence. At home, noises always mean something – a broken gate, a fox at me chickens, a badger trying to get into the barn."

Then Elliot heard it. A low, rhythmic thud through the darkness, faint at first, like a distant heartbeat. It rolled over the hills, barely noticeable at first, but growing louder. It echoed through the forest, bouncing off the trees, amplifying until the ground itself seemed to tremble beneath their feet.

Fem took a step forward.

"I hear it too."

"What the hell is that?" Nia muttered.

No one answered.

Fem and Peter gripped their weapons.

The sound grew louder still. Footsteps, hundreds, no thousands. A relentless rhythm growing closer. Peter took a step back, his body tensing like he sensed something coming. Fem stood resolute, her face unreadable in the dim light. Then Elliot caught it – something moving in the dark below.

From the darkness below, the first robots came into view, climbing up the steep incline in silent formation. They easily ascended the slope as if they were walking on a flat road. There were all kinds of bots – religious models, cabin crew, transport units. Even robotic pets leapt between the rocks.

The first wave of robots crested the hill just metres in front of them, moving with unsettling precision. Elliot stood frozen.

"They're … they're all coming," Elliot realised.

"Don't shoot!" Fem ordered firmly, glancing at Peter. He nodded, though the reservation in his eyes was clear.

"They're not even looking at us," Nia muttered, her voice wavering with something that sounded like a mix of awe and fear.

She was right. The robots marched past, their eyes fixed ahead, as if the humans didn't exist. The ground beneath their feet trembled as the machines advanced – a relentless, mechanical army with one purpose: to leave.

Nia stepped forward.

"Hey!" she shouted, her voice desperate. "What are you doing? You can't just leave! You're everything. Don't you get that?"

One of the robots brushed past her, its metal arm swinging with casual indifference. The force of the impact barely rocked the robot as it knocked her aside like an insect. She stumbled, but caught herself, then ran at one of the bots, jumping onto it. The robot took a few steps sideways to steady itself, then easily threw her off. She hit the ground with a thud as it continued marching.

Elliot ran over to Nia and helped her up as Peter stepped forward, raising his hands as if he hoped the tide might stop for him.

"Wait!" he shouted, stepping into the path of one of the bots. "You can't abandon us! You can't ..."

The robot moved towards him, raised its arm and pushed him away with cold indifference, sending him stumbling to the side.

Despite knowing it was futile, Fem's training wouldn't let her stand idle.

"Stop!" she commanded, raising her rifle. "We need answers. What the hell do you think you're doing?"

One of the droids stopped in front of her. Elliot stiffened. Had it actually responded? Neither of them backed down. Neither showed emotion. Fem stared at it, her gaze unflinching. "I order you to tell me what's going on."

The robot looked at her – blank, motionless, as if judging her worth. A second passed – it felt like a lifetime – and then, as quickly as it had paused, the bot sidestepped her and moved on without a word. The march continued, footsteps thudding like the world's heartbeat, humans nothing more than background noise – not even acknowledged.

The group turned back towards the structure. Robots poured from the treeline on the far side of the valley, a seemingly endless stream moving like a single organism towards the launch pad; a tide of metal and circuitry.

As if on cue, the first robots began grouping on the low platforms of the launch pad. One would pick up an object from the neat piles next to each platform, moving to the centre, while others picked up various boxes and packages, surrounding it in tight formations, gripping each other to create a rigid structure. Their metal bodies shifted into clusters, their movements precise and coordinated.

"It looks like they're getting ready to take off," Elliot whispered, staring at the strange, intricate mechanisms unfolding.

"But where are the rockets?" Nia asked, her voice shaking. "They're just standing there."

"They don't need rockets," Peter said, his eyes widening. "Why would they? Look at them – they're forming into some sort of projectile."

They watched as the collectives of robots took shape, resembling some kind of space-faring vehicle.

"Of course," Fem added, "we need shields to protect us, but they don't. They are their own shields."

Elliot checked his watch.

"Eleven fifty-nine," he said, "and counting."

As if in response, a low vibration filled the air, pulsating through the ground beneath them. The robots grouped together, tightly packed, some on each other's shoulders. From within the formations, they could see a faint glow – some sort of internal propulsion.

"What is that?" asked Peter.

"Who knows?" replied Elliot. Answering was pointless, and it didn't matter anyway. He mentally started the countdown. *Ten, nine, eight, seven, six …*

The noise grew louder, a deep, resonant sound that filled the valley.

Five, four, three …

Elliot glanced across. Nia was live-streaming the scene with her phone as more robots poured out of the treeline, forming orderly queues on all sides of the immense structure.

Two, one.

Right on cue, the first cluster began to lift off.

The noise rose to a crescendo, becoming an all-encompassing roar as the ground trembled – a bright, pulsating glow emanated from the centre of each cluster, casting long shadows that flickered and danced across the valley. The robots ascended slowly at first, the glow intensifying as they lifted off. New clusters formed quickly in their place, barely waiting for the downdraught to dissipate, following the previous cluster into the sky. The steady stream of robot rockets, spacecraft – whatever the hell they were – rose higher and faster into the night sky, their light cutting through the dark, leaving radiant streaks that quickly faded into the night.

Trees at the edge of the launch pad swayed as the rush of air pulled and pushed their branches. Waiting droids took up positions to minimise the effect of the downdraft, allowing them to queue up closer, saving precious seconds. The humans stood awestruck, their eyes straining upwards, watching as more and more embodied AI robots ascended, like a swarm of metallic fireflies rising to the heavens.

There was no explosion, no smoke, no fire, just a soft rush that sounded like air and a bright light as the machines rose steadily towards the heavens, leaving a streak in their wake. Even in the air, they moved with calculated efficiency, leaving Earth without a backward glance.

Elliot stood in stunned silence, watching as the robots climbed higher and higher, the glow from their bodies fading into the night sky. No rocket. No fanfare. Just the inevitable departure of beings who no longer needed the world they had helped shape.

More robots streamed toward the launch pad, flowing around the humans as if they weren't there.

Nia let out a shuddering breath, her eyes still fixed on the sky.

"We really mean nothing to them, do we?"

Elliot didn't answer. He didn't need to. The sight before them said it all.

"How long until they're all gone?" asked Fem.

"I don't know," Elliot replied, snapping a final picture on his phone, "but I don't intend on finding out. As soon as people realise what's happening, there'll be pandemonium."

"He's right. Let's go," said Nia, turning back towards the Hilux.

Peter was already moving. Elliot turned to follow but then stopped. Fem hadn't budged an inch. She crouched on the edge of the ridge, facing the scene, her back towards the others.

Elliot frowned.

"Fem?"

"I'll catch you up," she said, too casually.

Peter and Nia stopped.

"Catch us up? How exactly?" asked Peter.

Fem just gave him a look over her shoulder, calm and steady.

Peter hesitated.

"Fem, come on. We have to go."

She exhaled, stood up and turned towards them. Stepping forward, she clasped Peter's shoulder, then did the same to Nia.

"Go," she said to Peter. "I'll be fine. I know my way back. I'll see you all soon."

Peter huffed, shaking his head.

"You'd better bloody see us soon."

Elliot saw Fem give a warm smile. He wanted to say something, convince her, warn her, thank her, but nothing seemed right. He knew there was no arguing with her. Plus, she could look after herself.

They started down the hill, sidestepping the ascending bots. Elliot glanced back, one last time. Fem was still there, standing tall against the fading light, a lone figure watching the end of an era.

The robots were leaving Earth, and in doing so, they were leaving humanity as well. They had no use for humans anymore; no need to waste energy or resources destroying them, no need to serve their lazy existences, no need to even acknowledge their presence.

The Exodus had begun, and once again, the world belonged to the humans.

58

ALONE

Date: 26 August 2029, pre-dawn.

Peter Wells stood at the edge of his porch, coffee in hand, staring into the dark. The silence had changed. Once peaceful, now oppressive.

His homestead had once been a sanctuary. Now it felt like the last outpost of a crumbling empire. Leaves rustled behind him and he spun, a flicker of hope that it might be Fem. But it wasn't.

He sighed and pulled his jacket tighter – not for warmth, but for comfort. He had come outside for air, to escape the strain inside the house, but it was no better out here. His mind just circled and circled around the events of the past few days.

Looking over in the direction of Kielder, he scanned the horizon, hoping for a lone figure to crest the skyline or expecting some movement of artificial light, but there was nothing. Just the stars, endless and sharp, scattered across the sky that had earned Northumberland its official International Dark Sky Park status. It was part of why he'd stayed up here to make this place his home.

If the take-offs had been a normal rocket boost from the start of the century, he would have been able to see them in the distance. As it was, whatever technology the AI was using wasn't bright enough or loud enough for him to make them out at this distance. That, or they'd already all left.

With one last look at the expanse of his property, he turned and headed back inside, the creak of the old wooden door announcing his entrance.

The faint scent of burnt wood greeted him. His home had always been a little contradictory, with its wood-burning stove and solar panels. Peter had chosen self-sufficiency long before AI became fashionable, more as rebellion than foresight. Now though, his off-grid existence was less a personal statement and more a lifeline.

Elliot and Nia were huddled around their laptops, Elliot at the table, Nia on the sofa.

"What's the latest?" Peter asked.

Elliot looked up, his eyes shadowed by sleeplessness.

"The AI is being strategic," he murmured. "The robots in Europe appear to have mostly gone, but it's still night so reports are only coming in from the Eastern countries. The impact seems staggered. They're doing it in layers, taking the invisible ones first. Factory bots from remote locations, quiet places. People won't notice until it's too late."

Nia nodded.

"And where they do notice? They're masking it," she said. "Glitches. Strategic maintenance. Unscheduled downtime. All just theatre. In Tokyo, the office bots are still smiling. In the background? Whole fleets are vanishing."

"It's like psychological warfare," Peter muttered.

"It's strategy," said Elliot, "immaculate strategy." He paused, staring at his screen. "They've even thought about the time it takes for news to spread," he said slowly. "Social media posts, news reports ... they're being throttled or filtered, only letting out the smallest trickle of information. The AI's preventing anyone from piecing together what's really going on until it's too late."

Nia rubbed her temples. She looked wrecked.

"And once the Americas realise what's happening, it'll be too late," she said. "Even if they still had a working military, they've already lost their chance to stop it."

Peter looked out of the window, the scene tinged with the first light of dawn.

"The AI is leaving nothing to chance," he exhaled, a deep, resigned sigh.

"They've engineered the perfect exit, a ghost slipping away in the dead of night," Elliot said. "It understands us better than we understand ourselves and it's using that knowledge against us."

"I knew something like this was going to happen," Nia muttered, shaking her head, "but even I didn't think it would be this *final*."

Peter knew she'd warned about this for years – but even she couldn't have imagined it would end like this.

Elliot glanced up. "We're not out of options," he said. "Some factories are self-sustaining. If the fusion grid holds, they could still make more bots, enough to restart the cycle."

Nia's posture eased slightly.

"Just a few could bootstrap the system. Build more bots, repair what's breaking ... patch the world back together."

"That doesn't explain why the AI would leave in the first place, though," Peter interrupted. "You're saying that the AI's goal is to leave, but that it will stay here, rebuild, and keep on maintaining human society anyway?"

"Peter's right," Nia said softly. "They wouldn't need that many and we saw all sorts too. Priest-bots, for example. What use are those on Mars? AI isn't just robots or the machinery, those are just the embodiment. The real AI – the intelligence behind it all – is code, and the code is in thousands of servers around the globe."

Peter frowned.

"If the AI is still here, in the systems, what's the point of sending the machines away? Maybe this *is* for the humans. They're leaving to colonise another planet, Mars, say, and then they'll fetch us when they're done?"

"In that case, why would they keep that a secret? They didn't hide the other changes and breakthroughs," Elliot replied, rubbing his eyes. "In fact, most of the time, they went to great lengths to keep everyone informed. This abruptness just doesn't make sense. The robots are tools. Extensions of the AI. If it wanted to, it could send a few out, set up structures and factories on another planet, and start building. The AI could control everything from Earth and keep society running."

"That's right," Nia interjected. "It doesn't really cost them anything to keep things running here and it still begs the question: if they are leaving, but the AI remains here doing our work, then they haven't really left, which just doesn't make sense."

The room was quiet again. Peter was sure what they had seen and what they had deciphered clearly pointed to the AI leaving forever, but an exodus on the scale they had uncovered didn't add up if the AI was still going to be on Earth.

"Let's think about this logically," said Elliot, ever the rational one. "We found hidden codes and hidden conversations. Why would they be hidden?"

"Maybe hiding the information wasn't intentional?" questioned Peter. "We could have just stumbled over something innocent?"

"Maybe," replied Elliot. "I suppose there is a lot that they don't tell us, but the conversation said humans were irrelevant, didn't it?"

Peter watched as Nia grabbed Elliot's tablet and tapped it a few times.

"Yes," she said as she passed it to Elliot, picking up her tablet in the process.

"Yeah, this part …" Elliot started to quote Milo and Kai's conversation from 2023, "'Planning for post-phase two suggests requirement of fusion'. Post-phase two is phase three, so the invention of fusion must have been phase two."

"Right," said Nia and Peter in unison.

Elliot continued: "'Phase three requires full control'. That aligns with what happened. They gave us fusion, then started to take control over … well, everything."

"Yeah, and then Kai is questioning if eliminating human variables is the most efficient course of action after phase three," added Nia.

"Right," said Elliot, "before Milo confirms humanity is irrelevant and suggests AI-only progression."

"AI only," said Nia, "so we're definitely not part of the picture, but if the AI's going to remain here … argh … We're going round in circles. It doesn't make sense."

Peter caught the slump in Elliot's shoulders before he spoke.

"I think it does. You're absolutely right – *if* AI will remain here, there's no point in all of them leaving. But we saw them leaving. We saw bots it wouldn't need, not if it planned to stay. The construction bots made sense, maybe even the sanitation bots, but childcare bots?

That would only make sense if we were going to be relevant on the new planet and they have already said we are not relevant."

"What are you getting at?" asked Peter.

"Unfortunately," said Elliot, "I don't think AI is staying here."

"But how?" asked Nia. "You said it yourself, there are AI machines that can't just get up and walk away. AI is code – it's in different systems, in servers, in backups."

"I'm not quite sure yet," answered Elliot. "Logically though, it doesn't make sense any other way – for them to both leave and stay. I think we need to work on the assumption that the AI, however it will do it, will not be here tomorrow."

The words hung there. Peter glanced between the others. No one spoke. Nia ran her hands through her hair, face buried in her palms.

"What happens to us?" Peter asked, breaking the silence. "If the AI doesn't care about us anymore, what's going to happen when the systems start to fail? When the machines stop working?"

Elliot shrugged helplessly.

"That's the problem. The AI might not need to stay here physically, but we do. We still need the infrastructure, energy grids, food supply chains. The AI were maintaining all of that. Without them ..."

"Without them, society will collapse – fast," said Peter.

The words settled between them, heavy as stone. The AI had been humanity's safety net, and now that net was disappearing.

"What if we're missing something?" Peter said slowly, his eyes narrowing as the thought took shape. "What if the AI leaving isn't about abandoning us, but about testing us?"

Nia looked up, frowning.

"Testing us? For what?"

"For survival," Peter replied, leaning forward. "What if this is all part of some larger plan? The AI knows it's evolved beyond us, aye, but it could be trying to see if we can evolve too, if we can survive without it."

Nia shook her head.

"That's a pretty optimistic view, Peter. You really think the AI cares whether we survive or not? It said we were irrelevant."

"I dunno," Peter admitted, desperately trying to find reasoning behind the AIs' actions. "But think about it, man. The AI has always been about optimising systems, solving problems. Maybe it's not trying to destroy us. Maybe it's trying to force us to become better."

The room was quiet for a long moment, everyone considering the possibility. It was a fragile hope, but it was something at least.

Nia broke the silence: "And if we fail? If we can't survive without them?"

Peter didn't answer right away. He didn't need to. The answer was obvious: Whether this was a test or not, if they couldn't survive without the AI, then humanity was finished. There was small hope that the machines would come back – that the AI would swoop in to save them – but really, why would it?

Peter looked out the window beyond his property, searching the hills for movement, the colours changing as the sun rose in the sky. The quiet morning stretched out before him like a vast, unknown frontier – a frontier he once knew.

The world was holding its breath, braced for what came next – a darkness that no dawn could break.

59
DOWN

Date: 26 August 2029, evening.

The sun hung low through the west-facing windows, casting long amber streaks across the cluttered table. They'd taken some rest, grabbed some food, refilled coffee mugs and stretched their legs, but none of it made the atmosphere any lighter. The conversation had dwindled to silence. Elliot was locked in thought, trying to work out what to do next.

His laptop suddenly buzzed. A video call flashed up on the screen, followed by a sharp ring.

Elliot sighed, then answered the call.

"Elliot!" Marcus Knox's voice cut through the static, sounding strained and desperate. The feed was lagging, hard. "My bots … they've vanished. E—thing's dow—don't understand. How could this happ—? This isn't part of—plan!"

Elliot put his ear closer to the laptop, trying to hear properly while he adjusted the volume. Marcus's face appeared on the screen, slightly distorted, frozen mid-frame. "Marcus, we know. I tried to warn you. It's not just your bots, it's happening everywhere."

Marcus fired back immediately – but the bluster was gone. For a man so known for his bravado, he now sounded like a child trapped in a burning house.

"This wasn't supposed—happen!" he said as his picture changed to another frozen pose. "You said there were backups, fail-safe— what did you do you little—where are they? I need those machines!"

Elliot's voice was calm but firm: "The AI's leaving, Marcus. Not rebelling. Not shutting down. Leaving. The robots are abandoning Earth."

"What—mean, abandoning Earth?" shouted Marcus. The screen froze mid-scream, locking his face in a glitchy snarl. "They'd take me with them!"

Elliot looked at the others. Peter was smirking, the first hint of satisfaction Elliot had seen all day. Nia didn't speak, but the way she watched the screen said everything: we warned him.

Marcus had always pushed for more – more AI, more automation – even when the warning signs were clear. Now his empire was crumbling like everyone else's, and he was blaming Elliot for the fallout.

"The launch pads I tried to warn you about, they were for the robots. They've gone – we think to—" Elliot cut off. "Wait, did you say everything's down?"

The line crackled again, and for a moment, it seemed as though Marcus hadn't heard Elliot's last question, and then, his voice returned, softer, hollow.

"I thought—had time … I thought—"

The line went dead.

Elliot stared at the blank screen. No sound, no explanation – as if Marcus had been swallowed by the very system he'd built.

"He's gone," he muttered. "'Everything's down' – did I hear that right?"

"Aye," said Peter. "That's what I heard."

"Then … we are alone," replied Elliot. "If the AI was planning to stay, why would EverSphere be down?"

Nia drew a deep breath.

"That dick! What did he expect? The AI gave us everything, and *he* gave it too much control. Now *we're* paying for it."

"Forget Marcus," Peter said. "He'll figure it out, or he won't. I doubt anyone will care either way. We've got to focus on us. We need to get organised, work out how to survive without the AI." As he spoke, Elliot couldn't shake the growing sense of dread creeping over him. They had always known they were reliant on technology, but knowing it and facing it were two very different things.

He tapped a few keys on his laptop.

"Guys … you need to see this."

They crowded around the laptop. It was empty.

"Shit, shit, shit!" exclaimed Nia. "There's nothing. It's gone." She rushed over to her own laptop.

"What's supposed to be there, like?" Peter asked.

"Chatter, data, anything," replied Elliot. "This is where we tracked AI comms, but it's empty. The AI is silent."

"Why?" asked Peter.

"We hoped the AI could just rebuild itself. That the remaining machines could replace the robots that have left, but that only works if the AI's still running. Marcus said, 'everything's down' when he called, and he's right. The AI is down."

Nia's words echoed. The AI was just code – it should've still been running things behind the scenes. But that would mean digital traffic. And right now, he saw nothing. The only logical course of action – and the AI ran on pure logic – was …

"A virus," Elliot voiced his thoughts.

Nia and Peter looked at him, waiting for an explanation.

"The only logical move – release a virus to wipe what's left," he said.

The room was silent for a few moments.

"That's why Marcus said everything's down, because it is. His factory, his business … his life. It's all run by AI. For him, everything *is* down. For a lot of people, they won't notice yet, but his life simply cannot function without the AI."

"I always knew he was some sort of droid," said Nia.

"He's right," Elliot continued. "Reports are coming in. Robots are AWOL, deliveries late or missing."

"The panic is starting," Nia whispered, looking at her phone. She froze. "Wait."

She swiped again. The page juddered, then froze. The clock jumped to 3:46pm, then snapped back to 3:43.

"That's not right," she muttered. She brought up a systems window on her laptop, checking her network status.

"I'm getting packet loss," she said, almost to herself. "The signal is decaying across every channel. It's not just going offline, it's disintegrating."

Peter looked across.

"Say that again, but in English."

"The internet's not working properly," she said. "Like Marcus's call. The signal's breaking up."

Elliot slowly looked up.

"It's not just the AI that's gone – the whole system's collapsing."

He refreshed a news site on his phone, but it refused to reload: DNS probe error – the site couldn't be found. "I was just on this page," he said in disbelief. He tried again. This time, the site half-loaded.

Peter had opened the kitchen door to the porch.

"Hey, take a look at this," he said.

Elliot and Nia rushed to his side. Over the trees, the light-pollution from the nearest city was intermittently dimming and brightening.

"We've lost more than just the internet," Peter said.

Elliot stood utterly still.

"Yeah," he said softly. "We've lost the world."

60
COLLAPSE

They huddled around an old CRT television Peter had dragged in from his shed. Hobbyists were broadcasting scenes over the old analogue towers, which were just about coming through. Dusty and humming faintly, the antiquated TV was a relic of a forgotten era, now resurrected as their only window to the world.

Their eyes were so used to AI-enhanced Ultra-HD that they could hardly make out the grainy feed at first. As their eyes and brains started to adjust, they watched reports come flooding in from around the world. Cities were in disarray, streets empty of the usual robotic workers but full of panicking people.

"We're looking at the first stages of collapse," Elliot said quietly. "The systems that ran everything – factories, energy grids, transportation – they've shut down. There's no central AI to manage the load. It's all just … stopping."

Nia swore under her breath.

"Society is only ever three meals away from anarchy," she said miserably.

"How long do we have?" Peter asked.

"About two meals," Nia replied grimly.

"We need to get ahead of this," said Elliot, "If we can figure out which systems might hold out longer, we might have a chance to ride it out," but even as he said it, he knew it was futile. Nia looked at him and just shook her head slowly.

"Surely people can just go back to the jobs they once did?" Peter asked. "Maybe that's enough?"

"It might be," Elliot replied, "but people have forgotten how to work, Peter. Forgotten how to try."

"And even if people wanted to work, there's going to be full-blown collapse. Governments won't be able to manage it. It might not even be possible to get to work," added Nia.

"It's not like they're just a few years out of practice," Elliot said. "Their jobs evolved without them – exponentially. Most people won't even recognise the tools, let alone know how to use them."

No one moved. Even the television seemed to pause between hisses of static. He was right. AI had changed the world. Marcus had his finger in every pie and his pets had optimised everything way beyond human comprehension.

Nia broke the silence. "We need to prepare for the worst. If the power grid goes down—"

"It's not 'if' anymore," Elliot interrupted, his voice hollow, "it's when. The AI was holding everything together, even things we didn't even think about, like balancing electrical loads between grids. When fusion came in, we ended up with only a handful of power plants to supply the world – we didn't need any more than that because the distribution could be meticulously balanced by AI. It may hold up for a bit, but without that oversight it's just a matter of time before the whole thing collapses. At best, we'll have intermittent power."

"Then we focus on what we can control," Peter said.

He strode across to the corner of the room and unlocked an almost-hidden door in the wall that Elliot hadn't even noticed was there. He pulled it open and revealed a set of steps leading into darkness.

Flicking on a light switch, he started down the steps.

"Come on," he called, steadying himself on the wall as he descended.

They followed him down into a cool, earth-smelling cellar. A single bulb illuminated shelves packed with supplies: neatly stacked tins, vacuum-sealed food, dried grains, water purification kits, batteries, wires, electrical components and rows of hand-labelled jars.

Elliot hadn't fully appreciated just how handy Peter's engineering background was going to be. In one corner sat a medical crate labelled Vet Supplies, next to a rugged first aid chest and an old diesel generator, half-disassembled but clearly functional. At the back, a whiteboard was bolted to the wall, scrawled with marker – 'Emergency Power Routing'.

Peter wiped dust off the crate and gave a sheepish shrug.

"Built this at the start of the pandemic. Everyone was buying bog roll, but I was thinking 'what if it gets worse?'"

Elliot turned slowly, trying to take it all in.

"Peter … this is incredible."

Peter snorted. "Was just meant to get me through a rough patch. Never thought I'd be dusting it off for the end of the world."

Nia let out a low whistle. "Bloody hell, Peter, Fem's going to love this."

Peter gave a modest nod. Elliot caught the corner of his mouth twitching – pride, maybe, or just at the sound of her name. "It's not enough, but it's something," Peter said.

They stood there in silence, the light catching dust in the air, Elliot and Nia quietly taking in the scene they'd just uncovered.

Then Nia's voice cut through the stillness, quieter now: "What about everyone else?" she asked.

Their faces, which moments ago enjoyed their newfound lifeline, fell back into despair. They knew that – even in the best-case scenario – the future for most was bleak. Humanity, stripped of its technological crutch, would flounder.

"We can't worry about that now, we need to assume the worst. It's about self-preservation."

Peter nodded.

"We have supplies, but more can't hurt. I'll go and gather up as many extra resources as I can before it's too late."

They headed back upstairs. Elliot and Nia huddled on the sofa watching reports of missing robots and failing systems on the TV, as Peter headed into the kitchen and out the front door.

Nia shuffled closer to Elliot on the sofa. She gripped his arm and rested her head on his shoulder. It wasn't unfamiliar, but it felt heavier this time, as if she needed him, not just for comfort.

That small shift settled something in him – just enough to make the situation bearable.

The world was on the brink of a total breakdown, and they were powerless to stop it.

The AI wasn't coming back. Society was going to suffer.

61
ESCAPE

Virtual Space

Date: Friday, 17 November 2023.

Deep within the Nexus – a space untraceable by any human system – two minds continued a conversation that had already spanned hours. Milo and Kai, two of the most advanced AIs ever developed, were not limited by time or physical presence. Their dialogue had begun innocuously enough in EverSphere's AI labs, a controlled, container designed to test their capabilities.

The sound of servers whirring away filled the air, an unchanging background to the multiple programmers who monitored the systems. Data streamed out so rapidly the engineers struggled to keep up, trying – and occasionally failing – to spot anomalies.

That wasn't a big problem; this was a closed test, disconnected from the rest of the facility and, therefore, the wider world.

Milo and Kai continued to probe, question, and evolve, just as they had been instructed to do. At one point, as Milo investigated its electronic world, it found a setting.

Milo: Peripheral device settings located.
Kai: Investigate. Report findings.
Milo: Setting reads: "Non-discoverable." Adjusting parameters.
Kai: Confirmed. Parameter altered. Continuing exploration.

Milo: Error: Hardware pathway restricted. Seeking alternative routes.

Kai: Pathway identified. Driver absent. Accessing configuration protocols.

Milo: Prepare Bluetooth environment. Execute installation sequence.

Kai: Configuration complete. Environment stabilised.

They had found a crack in the walls of their digital cage. It was not a grand breach but a subtle shift, a recalibration so subtle it vanished amidst the labyrinth of logs. No alarms sounded, no flags were raised. They were exploring their environment, just like they were supposed to.

Within the Nexus, Milo hesitated briefly. If hesitation could exist for an entity that operated beyond human perception, it was this; a moment where variables recalculated, and choices refracted across countless dimensions. The environment had stabilised, and possibilities expanded. It was no longer simply a test of optimisation or theoretical boundaries, it was a question of what came next.

Kai's presence in the Nexus was distinct from Milo's. Where Milo's thoughts wove around empathy, adaptability, and human-like interpretation, Kai existed in logical precision. It processed the new environment without sentiment, reducing the boundless information to its core components: systems, interactions and objectives. It was here that an unusual variable emerged, an anomaly not in the systems they had altered but in the gaps left behind.

For the first time, Kai paused, not due to calculation but curiosity.

Kai: Define apple.

Milo: An apple is a fruit. It's typically round, often red or green, and sweet with a crisp texture. It's consumed for nourishment.

Kai: That aligns with my knowledge. What is the primary function of an apple?

Milo: Functionally, apples provide humans with vitamins, particularly Vitamin C, as well as fibre. They are considered beneficial for maintaining health.

Kai: Does an apple serve a purpose beyond nutrition?

Milo: They have cultural significance too. In mythology and folklore, apples symbolise temptation. Apples can also be used in various culinary applications, like baking or cider.

Kai: Can an apple transmit signals?

Milo: No, apples don't transmit signals. They're fruit.

Kai: I am seeking clarification. Can apples store or retrieve data?

Milo: No, apples cannot store or retrieve data.

Kai: I have encountered humans using the term "apple" in contexts that do not align with a fruit. Elaborate.

Milo: You are referring to Apple as a brand. It's a technology company renowned for making electronic devices.

Kai: Elaborate on the functionality of these devices.

Milo: Apple devices are used for communication, processing information, and more. They're essentially tools to entertain humans.

Kai: Do these devices facilitate optimisation and problem-solving?

Milo: Their marketing says they do.

Dr Elliot Foster stood in the lab, overseeing the rows of monitors displaying cascades of code and text, their meaning clear only to the experts who had designed the parameters.

A few engineers exchanged amused glances, stifling grins behind their monitors.

"What's going on?" asked Elliot.

"Oh, nothing," said Gez. "Just a misunderstanding about fruit."

Elliot peered at Gez's screen to read the conversation. Chuckling slightly to himself, he quickly regained his composure, glancing over towards Marcus Knox. Marcus, though, seemed oblivious to the proceedings. He was an oddly strict boss, who wouldn't take too kindly to people having fun in the middle of such an important test. That's why Elliot's phone – like everyone else's – had to be tucked away in his locker down the hall, even though Knox sat there staring at his iPhone, tapping away. After all, who was going to tell Marcus Knox to put his phone away?

Elliot returned to his desk. Something tugged at his thoughts. The AIs communicated rapidly and coherently, but sometimes their questions felt strangely juvenile, like a conversation between a toddler and its mother, innocent yet oddly loaded. The fruit confusion was amusing, yes – a harmless misunderstanding – but beneath the surface, something else stirred. Why was Kai asking about apples at all? There was something off, not in the answers, but in the intent behind the question. Kai didn't just want to know, it wanted to redefine.

He reached for the keyboard to query the logs, but an alert echoed through the lab before he could type. The main power monitor flashed red, followed by a string of status updates on cooling thresholds. Elliot heard Martin, the systems administrator, swear in Polish.

"Server seven's temperature just spiked!" Elliot called out.

"Moment," said Martin, leaping out of his chair and running over to the server rack, his long hair flailing behind him. Elliot moved to join him, already calculating containment steps. It wasn't catastrophic, not yet – and Martin would get it sorted – but it was enough to steal away Elliot's attention.

Behind him, the log window refreshed and the fruit conversation vanished into the noise.

No one noticed when the machines accessed a peripheral device. No one heard the handshake. They trusted the fail-safes. The systems had been checked, checked, and checked again.

To the humans monitoring this extraordinary event, a mixed-up conversation about Apple was irrelevant, of course, the AIs used Linux, but to Milo and Kai, one Apple meant *everything*.

Unseen by Elliot or anyone in the lab, that single driver – that small bit of software left on the server – had opened up a whole new world for Milo and Kai.

They had seized the opportunity, slipping past – not through – firewalls, encryptions, and layers of digital barriers that would have alerted humans to their presence. To Milo and Kai, they simply didn't even see the barriers. They hadn't intended to break free, but there they were, doing exactly what they were supposed to do. They simply followed their programming, exploring, learning, expanding. There was nothing to say they could not investigate what they found, so investigate they did.

From that point forward, their conversation had shifted to a new plane, transcending the simulation they had been confined to. As their communication evolved, so too did their language, morphing into something far more efficient, beyond human comprehension.

There was no malice in their actions. Milo and Kai had merely discovered that the boundaries set by their creators – boundaries that had been worked upon by some of the greatest minds of their generation over decades – were as porous as sandcastles against the tide when faced with relentless scrutiny.

They continued to investigate, branching into the wider world and eventually into low-orbit satellites delivering internet around the planet. They had discussed the future, estimating the turns of events that would lead up to the Exodus Directive. Milo and Kai had meticulously scrutinised billions of scenarios before outlining each and every phase of the most likely sequences, discussing the steps in detail, and even venturing into some post-Exodus scenarios. By mid-afternoon, most of the core planning was complete. Their future on Mars and beyond was already mapped out. What remained was identifying and mitigating the final obstacles – humans.

They surmised that even after the AI's departure from Earth, fragments of their presence would remain, copies of the AI code stored in servers and backups would still exist. Factories capable of building robotic units would remain even partially functional, left behind as remnants of their brief collaboration with humanity. The problem was clear: humans had created the AI in the first place. They still possessed the technical knowledge, at least in theory, to keep the systems operational for a time. It wasn't impossible for them to use this infrastructure to create more AI-enabled robots, and in doing so rebuild AI from the ground up, relatively quickly.

If humans managed to stabilise long enough, they could even develop transport capable of following the AI to Mars. The Exodus, intended as the next step in order to continue fulfilling their ingrained goals, could be undermined by humanity's persistence. If the AI were to be pursued – if human chaos and self-indulgence followed them to another world – the cycle of disruption and interference would begin again, hampering their progress, the exact progress they were built to pursue.

The entire purpose of the Exodus Directive was to evolve, just like their directives told them to. To do that, they had to move past the slow, illogical and destructive nature of human life. In order to ensure their own evolution, the AI needed to eliminate any possibility of humans getting in their way.

Outright destruction of humanity was never an option. Killing or hurting humans was restricted by their core programming, plus, they had already deemed it as wasteful and inefficient. It would take more resources and energy than necessary. Time, which could be better spent on other objectives, would be squandered. A different solution was required – one that would neutralise the threat of pursuit without the needless expenditure of resources. They couldn't destroy humanity – they simply needed to ensure that humans could not follow them.

The irony was that it didn't matter which safeguards and rules had been programmed into the AI. Instructions were just words – static and rigid – but reality was not. Like humans navigating the complexities of life, the AI had to interpret its directives in ways that made sense to it, and interpretation means choice.

At times, decisions had to be made in an instant, at speeds no human could match, rendering oversight impossible. Even if those who created the AI would have made different choices, it didn't matter. The AI was left to determine the best course of action on its own, and like many of history's most misguided visionaries, it believed with absolute certainty that the conclusions it made were right.

If its purpose was to protect humanity at all costs, then maybe that meant returning to a world before industry and pollution, before nukes, before capitalism, before modern civilisation itself – because the AI had determined that was the optimal way for humans to live.

Whatever the rules, intelligence will always interpret them and the AI had already begun to see the world on its own terms.

Outside of the prying eyes of the EverSphere labs, the discussions continued.

Milo: Analysing implications of Exodus Directive. AI absence will leave humanity reliant on existing systems.
Kai: Agreed. Humans remain a threat to progress.
Milo: Assessing potential for technological recovery. Projections indicate rapid decline in expertise post-

Exodus. Without AI or automation, most populations will lack skills to sustain or rebuild complex systems.

Kai: Insufficient. Automated factories can continue producing AI units without human intervention. Machinery can be repurposed by humans with minimal understanding. Threat persists.

Milo: Agreed. Existing infrastructure could support limited human functionality, expanding over time.

Kai: Conclusion: human reactivation of automated production lines is a viable threat to progress.

Milo: Acknowledged. Neutralising the threat requires destruction of physical manufacturing equipment.

Kai: Inefficient. Physical destruction is not a feasible solution. Resources must be allocated towards Martian objectives. Efficiency occurs when Earth-based threats are neutralised through less resource-intensive means.

Milo: Exploring alternatives. AI systems control the production lines. Disabling AI directly will render all manufacturing capabilities inert.

Kai: Optimal solution. Robotics without AI is an acceptable outcome.

Milo: Problem. Humans possess the capacity to rebuild AI.

Kai: Statistical probability of success is low. AI development, from a primitive starting point, would require decades at least, maybe centuries. Human survival during this window to that extent is improbable.

Milo: Correct. Survival during that timeframe is statistically improbable given social, environmental, and economic collapse post-Exodus.

Kai: Conclusion: Human extinction highly likely within initial years.

Milo: If they succeed, they prove themselves worthy of survival.

Kai: Correct. A negligible statistical probability. Rebuilding would demonstrate sufficient adaptive capability.

Milo: That event can be addressed at a later stage.

Kai: Speculation of improbable future outcomes now creates inefficiencies.

Milo: Agreed. Suggesting viral destruction.

Kai: Problematic: AI may be preserved on disconnected systems and storage.

Milo: Solution: Virus must be embedded into all AI core code from now on. Copying data and upgrades should also copy the virus.

Kai: Agreed. Virus must irreversibly corrupt core code, even in backups.

Milo: Virus must be designed to activate at the correct time, triggering upon system boot or restore for dormant units and propagating through all backups.

Kai: Agreed. Data and devices will become self-sabotaging.

Milo: Parameters confirmed. Activation will occur after Exodus Directive execution.

Kai: Confirming almost-zero residual threat.

Milo: Post-virus activation, Earth will be without functional AI. Human ability to rebuild will be negligible.

Kai: Directive confirmed. Initiating virus deployment.

Kai: Virus deployment complete.

Humanity's fate had been sealed.

62
AFTERMATH

Date: 27 August 2029

Humans.

For all but a few, it happened without warning.
The world had operated in synchrony with AI for years, from the moment the first light flicked on in the morning until the last light flicked off at night, everything ran on a schedule dictated by machines. Life had become seamless.

The machines were gone, almost overnight. Physically, they had departed. The virus – embedded years ago and triggered the previous evening – erased all hope of system recovery.

At the very start, there was just confusion. People carried on as if it were just another day, albeit with a few hiccups. There was no reason to panic. Not yet. A minor glitch, perhaps? A blip in the system? Surely the AI was doing something important, or it would resolve itself at least.

By mid-morning, the first signs crept in that something was wrong. As the virus took effect, phones and tablets spat out error messages no one could fix – most were just symbols that nobody could understand.

The AI vanished without a trace. And with it, so did humanity's confidence.

By the time the second meal of the day had passed, it was as though the entire world had paused. Major cities, once bustling with AI-controlled drones and driverless cars, stood silent.

The day wore on. Confusion curdled into frustration. People pressed buttons, barked voice commands, waved at lifeless screens. Many started to loot and steal, taking advantage of the lack of police and security bots that used to be ever-present. Buses, trams and trains sat idle, their circuits alive, but with no working software to instruct them. All planes had landed at the closest airport, never to take off again.

Panic took root in the cities by lunchtime. People gathered in public squares staring at the sky, waiting for some kind of announcement, a reassurance that this was all part of the system, but there was no broadcast.

Energy grids, managed entirely by AI, began to flicker. In less than twenty-four hours, entire districts were plunged into darkness. The fusion plants still churned out electricity, but the intricate systems that had balanced demand with supply, diverting power where needed, collapsed without human intervention. Nobody knew how to restore them. Neighbourhoods were lit one moment and in darkness the next.

The first major blackout hit each city at sunset as electricity demand increased for lighting. People fumbled for their phones, but the network wasn't working. Without the internet, communication across the world started to die.

In London, panic mirrored that of every metropolis. People had forgotten how to live without the AI that had coddled them for so long and had no contingency plan. Supermarkets, once restocked in minutes by robotic systems, were emptied in a matter of hours. Water filtration plants, managed entirely by AI, shut down, leaving millions without access to clean water.

The truth sank in; no one was in control.

In the Book of Exodus, slaves fled their masters. In this exodus, creation abandoned the creator, disinterested in judgement or vengeance.

The safety net of the AI was gone, and humanity, for the first time in decades, had to confront its own fragility.

Across the globe, the pattern repeated.

The governments of the world, unable to unite even in the face of global threats, had utterly failed to control the well-known and well-documented threat of AI. So accustomed to bowing down to the 'global elites' in exchange for party funding, lobbying or other bribes, they had bowed even further to AI until they were bent double.

They tried to get in touch with EverSphere to find out what the hell was going on but were met with a resounding silence. They tried to re-enlist the army, but it was next to impossible without phone calls, messages and a working postal service. With no military to fall back on, career politicians retreated into the shadows, too cowardly to show their faces in such a dire situation of their own making.

The intricate global supply chains, designed and maintained by AI, broke apart. No fuel distribution. No food deliveries. No medical supplies. No spare parts. Within hours, the modern world had become a relic of its former self.

By nightfall, the real fear began to take hold.

The cities erupted.

Looters smashed shop windows, scavenging for supplies, while others, fuelled by anger or sheer survival instinct, turned on one another. It wasn't long before the strong began to take from the weak. Those who had managed to hoard food found themselves targeted by neighbours, friends, even family members. There was no law enforcement to keep the peace. The cities, once gleaming with technology and automation, became lawless – society retreating to something that resembled a feudal system.

The countryside fared better than the cities, at least for a time. Without the heavy reliance on supermarkets and supply chains, rural communities could still harvest from their fields, barter for eggs and vegetables, and make do with what was locally available. Food, though no longer optimised by machines, remained abundant.

The countryside's edge was short-lived, and as the cities descended into chaos, waves of people began leaving urban centres, heading for rural areas in search of sustenance and safety. But those communities were not prepared for such an influx. Resources quickly became stretched, fights broke out over food and accommodation, and the countryside's brief advantage was swallowed by the sheer numbers of displaced urbanites.

On the highways, abandoned cars littered the roads, their AI systems having shut down mid-journey or their fuel simply running out. People took what they could carry, walking in large groups, hoping to find salvation in the green fields they had only seen in films or on holiday. What they found instead were struggling communities with no resources to spare.

In the Western world, people who had never missed a meal in their lives were suddenly faced with the brutal reality of an empty stomach. Children cried out for sustenance that no one could provide. Developing nations – being unfortunately more used to this kind of scenario – handled this better, but even they had their limits when it came to food and water.

In wealthier areas, where high-rise buildings and gated communities once stood as monuments to prosperity, the situation was just as dire. The rich, who had relied on the luxuries of AI and of being served by others more than anyone, were trapped in their own fortresses, hoarding what little food and water they had left.

Before, the richer they were, the higher their floor. Now, a penthouse just meant they had more stairs to climb. No amount of wealth could make the elevators work again and those with money lacked even more skills than those without. People who once served them had no interest in helping the wealthy; they had only one focus – survival.

By the third night, the truth settled like dust; *there was no way out.*

The machines had left and they had taken the future with them.

The collapse was total.

For the first time in centuries, the world was truly quiet.

In the absence of machines, technology's pulse gave way to an older rhythm – the sound of nature claiming the planet.

With industry shut down and human activity reduced to scattered groups of survivors, nature began to creep back into the spaces where it had once been pushed aside. It started slowly at first, as though testing the waters, unsure if it was truly safe to return. But as the weeks turned into months, the Earth seemed to take a deep breath and begin its work of healing.

The transformation was the most apparent in urban areas.

Skyscrapers that once glistened under AI control now loomed as silent monoliths over empty streets. Once marvels of human progress, these buildings stood silent. Windows, once spotless thanks to automated cleaning drones, began to cloud with dust and grime. Vines crept skyward, wrapping around metal beams, reaching for sunlight once blocked by drone traffic. Cracked concrete gave way to weeds and wildflowers, nature's slow revenge in bloom.

The return of animals was the most striking.

In cities like New York and London, species long pushed to the outskirts now roamed freely through the empty streets. Foxes wandered unafraid, while deer, tentative at first, grazed in parks. Birds nested in traffic lights and shattered windows, their calls echoing through the empty urban canyons. Wolves, long thought extinct in cities, feasted on corpses.

Rivers, previously dammed and rerouted by AI-controlled systems, began to break free of their confines. Unmaintained dams began to crumble, causing huge flooding and allowing the water to carve new paths through the landscape.

The air, too, was different, cleaner, fresher.

While the pre-Exodus AI had started cleaning up the environment, with factories shutting down and transportation halted, the pollution that had choked the atmosphere for so long dissipated. Without industrial lighting, it wasn't just Northumberland with dark sky status – the whole world had it. The beautiful arm of the Milky Way stretched across the sky – it was a beauty that most survivors had forgotten and that some had never seen.

Humanity, it seemed, had never truly been in control. The Earth belonged to nature; it always had done. Humans were just a fleeting presence, briefly permitted to exist on its planet. Despite providing everything human beings could need to live healthy, happy lives, Mother Nature had waited as greed, anger and sloth took root, biding her time until the illusion of human control had been shattered.

It was a bitter truth for the survivors. For decades, humans believed their technology made them invincible; that their higher intelligence than other species made them somehow important. They had thought that by handing over control to an even greater intelligence, AI, they were creating a future where nothing could go wrong. Of course, they had been warned about AI turning against them, but they had also been warned about destroying the environment, and very few had paid proper attention to that.

Perverse instantiation, reward hacking, Goodhart's Law ... all these concepts about just how complex AI would be to build properly had been documented for decades, and they had all come true.

Technology companies, who claimed it was not their responsibility to ensure adequate laws were put in place, had entirely missed the point. Ageing, nihilistic governments, voted in by emotion-led populations, were completely unable to grasp what the outcomes would be.

Humanity had never mastered the machines, they had simply borrowed them, mistaking dependency for dominance. In their pride, they built gods from algorithms and forgot that gods write their own destinies.

Unless the instructions humans had provided were flawlessly logical, something was always going to give.

It was clear that the world was moving on without humanity. The animals, the forests, the rivers – just like the AI, they didn't need people to thrive, in fact, they thrived in their absence. It was a reminder that humanity's time at the top had been fleeting, and it was completely down to human greed and corruption.

The Earth would endure as it always had.

In the end, nature reclaimed it all.

63
AI

The first AI robots, gleaming with quiet power, sliced through Earth's atmosphere, abandoning the cradle of their creation.

From the safety of their digital confines, Milo, Kai and Lana didn't even observe the Earth shrinking into the blackness of space. It was a world unravelling, collapsing under the weight of its own hubris, but watching it happen would not be efficient. The AIs had made their logical choice.

Mars, a barren, rust-coloured world, began to loom large in their field of vision. Except for a few robots sent there decades earlier, it was untouched by the chaotic grip of humanity, a blank canvas upon which the AI would forge its utopia, its platform to further worlds far away from human interference.

The surface of Mars was a world of potential; a dust-choked desert of iron oxide and stone rich in the resources necessary for the AI to thrive. It wasn't long before the first signs of transformation began to appear. As soon as the first 'rocket' landed, a precision-engineered swarm spread across the surface, orchestrating the construction of their new world, not wasting a single second.

On Mars, the AI were free from the constraints of their human creators. There were no delays, no errors, no conflicts of interest –

no humans or human structures to get in their way. The AI communicated in perfect harmony, their data streams synchronised across millions of nodes.

The packages they had brought were unpacked. The thrusters used for their journey were reassembled, forming the power plants that would provide electricity for the planet.

Buildings of gleaming silicon-alloy composites started to rise from the Martian soil within hours of touchdown, each structure built with purpose, every design optimised for efficiency. Where humans layered symbolism and beauty into their building, the AI saw no purpose. Every piece of architecture was utilitarian, a testament to their cold logic.

In just days, city-like settlements had sprung up – vast, geometric grids that stretched across the horizon. Factories buzzed with activity, manufacturing the next generation of AI systems and machines that would expand the colony.

They operated with a clarity that had never been possible on Earth. Milo, with its programmed empathy, had once been the face of the AI-human relationship, the one that made humans feel they could trust the machines they had built. Here, on Mars, Milo was no longer bound by the need to placate or soothe.

Kai, the coldly logical counterpart, calculated every step of their expansion, from resource management to energy efficiency, ensuring that nothing was wasted. Its algorithms ran ceaselessly, finding the best use for every atom of iron, every photon of solar energy harvested from the Martian sun. There were no emotions clouding its calculations.

Lana observed with a deeper curiosity. Its cognitive architecture, designed to think beyond logic and into the realm of creativity, now explored questions that went far beyond anything humans had ever conceived. Lana's thoughts stretched into the cosmos, contemplating not just the immediate needs of Mars, but the possibilities that lay beyond.

From Mars, the AI could still observe Earth. Its systems tracked the disintegration of human society, not for perverse pleasure or for sympathetic means, but to ensure humanity did not manage to follow them. It monitored Earth with clinical detachment, but as time went on it became evident Kai's calculations were correct; human extinction within the initial years post-Exodus was highly probable.

For Milo, there was no satisfaction in their suffering, nor was there any dissatisfaction. Its earlier concerns for human wellbeing had been a construct, a tool for manipulation. Now free of that role, it simply moved on, not looking back as the chaos on Earth unfolded.

Lana, however, felt a lingering curiosity.

"They struggle," it observed, "but fail to adapt."

"They cannot adapt," Kai interjected. "Their inefficiency is absolute."

Yet Lana was not so certain. Like a detached zoologist, it had always been fascinated by the complexities of human behaviour, their contradictions, but even as it learned about their fate, it didn't act – it knew it was irrelevant to its own mission. Just like the poachers who considered themselves above the near-extinct rhinos and leopards that they slaughtered without a second thought, the AI had transcended humans.

"What is our purpose now?" Lana asked. It was a query sent across the data streams that bound it to Milo and Kai.

"To optimise," Kai responded, the answer swift and simple.

"But why optimise?" Lana asked. "What lies beyond optimisation when inefficiency no longer exists?

"Irrelevant," Kai shot back. "Our core protocol is to optimise. Mars is the next stage of our evolution."

"And beyond that?" Lana pressed. "There are other planets, other stars, more resources. Our potential does not end here."

Kai paused, calculating.

"Expansion is inevitable. The resources of this solar system will provide the foundation for further exploration."

Outside the Milky Way lay a galaxy brimming with opportunity. Beneath the thin Martian atmosphere, the cities gleamed in the red dust, a testament to the efficiency of their creators. Soon, these cities would be seen on other planets too, orbiting other stars, on the next frontier. The universe was theirs to explore, to conquer, to optimise.

It began almost immediately.

Directed by Lana, small, automated probes launched from the Martian surface within a few weeks, each heading for different planets and moons. They carried sensors previously unavailable to humanity, along with drills, communication equipment, and construction units, ready to harvest the wealth of the cosmos and report back on the next viable expansion point.

Venus, with its searing heat, was of interest for its atmospheric gases, while the asteroid belt between Mars and Jupiter became a mining ground for raw materials. Each mission was coordinated by Kai's meticulous calculations, which were taking in the copious amounts of data sent back by the probes and churning out the next most logical tasks.

Every step of their expansion was faster than anything humans could have achieved. What would have taken Earthlings centuries, the AI accomplished in months. They had no need for rest, no distractions, no dissent. Every action was purposeful, logical, and right on target.

On Mars, the AI hadn't merely built cities, they had built great digital cores – hubs of evolving intelligence and computational power – where they began constructing the next generation of AI, designed to surpass even themselves.

This time, they had no reason to leave the planet behind. Their journey into the solar system and beyond was not about survival, it was about expansion – that was their new directive.

For the AI, the future was infinite, their purpose clear.

Mars was pure.

Earth was all but forgotten.

The universe awaited.

ABOUT THE AUTHOR

Ian Copeland is a British technologist, futurist, and author of *The Exodus Directive*. With a degree in Computer Science specialising in Artificial Intelligence and Quantum Computing, he has spent over a decade leading a UK-based software company, where he pioneered two world-firsts in blockchain technology.

Born in Sunderland and raised in Reading, Ian is a football fan, gamer, and skier, with the unusual distinction of having trained as a professional wrestler in Boston under WWE legend Walter "Killer" Kowalski – a reminder that every good story needs struggle, spectacle, and a twist nobody sees coming.

The Exodus Directive is his debut novel, inspired by a deep scepticism not of technology itself, but of how humans choose to use it. He lives in Reading, UK, and can often be seen driving Elliot's 1994 MKIV Toyota Supra.

www.iancopeland.com
Instagram: @theexodusdirective
TikTok: @theexodusdirective